A MISSING LINK IN CYBERNETICS

International Federation for Systems Research
International Series on Systems Science and Engineering

IFSR was established "to stimulate all activities associated with the scientific study of systems and to coordinate such activities at international level." The aim of this series is to stimulate publication of high-quality monographs and textbooks on various topics of systems science and engineering. This series complements the Federation's other publications.

A Continuation Order Plan is available for this series. A continuation order will bring delivery of each new volume immediately upon publication. Volumes are billed only upon actual shipment. For further information please contact the publisher.

Volmes 1-6 were published by Pergamon Press.

Alex M. Andrew

A MISSING LINK
IN CYBERNETICS

Logic and Continuity

 Springer

Alex M. Andrew
95 Finch Road
Earley, Reading
United Kingdom RG6 7JX
alexandrew@tiscali.co.uk

Series Editor:
George J. Klir
Thomas J. Watson School of Engineering
    and Applied Sciences
Department of Systems Science
    and Industrial Engineering
Binghamton University
Binghamton, NY 13902
U.S.A.

ISBN 978-1-4419-2584-8   e-ISBN 978-0-387-75164-1
DOI 10.1007/978-0-387-75164-1

Mathematics Subject Classification (2000): 93-02

Printed on acid-free paper

springer.com

Preface

In this book I argue that a reason for the limited success of various studies under the general heading of cybernetics is failure to appreciate the importance of continuity, in a simple metrical sense of the term. It is with particular, but certainly not exclusive, reference to the Artificial Intelligence (AI) effort that the shortcomings of established approaches are most easily seen. One reason for the relative failure of attempts to analyse and model intelligence is the customary assumption that the processing of continuous variables and the manipulation of discrete concepts should be considered separately, frequently with the assumption that continuous processing plays no part in thought. There is much evidence to the contrary including the observation that the remarkable ability of people and animals to learn from experience finds similar expression in tasks of both discrete and continuous nature and in tasks that require intimate mixing of the two. Such tasks include everyday voluntary movement while preserving balance and posture, with competitive games and athletics offering extreme examples.

Continuous measures enter into many tasks that are usually presented as discrete. In tasks of pattern recognition, for example, there is often a continuous measure of the similarity of an imposed pattern to each of a set of paradigms, of which the most similar is selected. The importance of continuity is also indicated by the fact that adjectives and adverbs in everyday verbal communication have comparative and superlative forms.

Primitive organisms are more obviously dependent on continuous processing than are higher animals, though at all levels life depends on complex regulatory processes having continuous character. The suggestion here is that continuous processing should be seen as more primitive, with concept-based thought evolved from it. It is only possible to speculate about the mechanism of the evolutionary development, but a plausible theory is advanced. Because evolution does not erase traces of earlier forms, this may shed light on little-understood subconscious processes that underlie concept-based thought.

I realised the need to combine continuous and concept-based processing in connection with a project to make an automatic controller with learning capability for an industrial plant. Since then, the same need has been acknowledged and to some degree satisfied in many contexts by the advent and vast elaboration of fuzzy theory. From the evolutionary point of view this is wrong, as it amounts to accepting the concept-based or "logical" part of the process as primary and grafting continuity onto it. In many applications of fuzzy theory, parameters of truth functions are chosen by a designer and retained if they prove successful, or modified otherwise. This means that the fuzzy controller relies on the collaboration of a human acting as a continuous adaptive controller.

Continuous measures come to be associated with intelligent information processing in subtle ways, as acknowledged in Marvin Minsky's reference to measures of similarity, or *heuristic connection*, between problems, and in the associated *basic learning heuristic* of Minsky and Selfridge. The latter requires measures of similarity between situations and between responses. This reappearance of continuity supports a potentially valuable view of intelligence as having fractal nature, with structures at a complex level, interpreted in terms of measures of this subtle kind, mirroring others at a simpler level.

Criteria of similarity of situation and response are fundamental to any process of learning or adaptation, including biological evolution. The success of any process of adaptation depends on representation of the evolving structure in a way that allows rather major change, probably in small steps, but without too great a chance of complete disruption. It is difficult to see how this might be achieved other than by means that can be decomposed into nested processes of "adapting to adapt" and probably "adapting to adapt to adapt" and so on. The viewpoint implicit in early discussions by Minsky and Selfridge and expanded here may therefore have deep implications for biology.

Because the brain is the supreme example, adaptation and learning are frequently considered in terms of neural nets, and a chapter is devoted to this, with particular reference to a principle that I termed *significance feedback* in an early report. The later principle of *backpropagation of errors* is a special case and is the basis of most applications of artificial neural nets, a field that has flourished in recent decades. I argue that other forms of *significance feedback* also merit attention.

I have used the term "cybernetics" in my title, and in the first chapter I have tried to explain what I understand by the term. At the present time its interpretation has become surprisingly controversial, with some writers claiming that cybernetics and AI have nothing in common. Certainly there is increasing emphasis on sociology (communication and control *between* people and groups) often to the exclusion of the study of communication and control *in* the animal and the machine. The latter was specified in Wiener's original title and if taken literally places primary emphasis on neurophysiology and computing, though it was understood from the start that the topic was interdisciplinary with wide ramifications. Its aim, with this emphasis, has been slightly paraphrased in reference to McCulloch as that of "understanding man's understanding", in a context that implies essentially mechanistic explanation, and for this the achievement of forms of artificial intelligence must be the ultimate "proof of the pudding".

Another general point that should be made is in connection with the small amount of rather elementary mathematics that I have included. I have great respect for the many highly mathematical treatments that bear on systems theory but believe they are not relevant here because they provide tools for use by sophisticated people whose thought processes in applying them have gone well beyond the origins of concept-based reasoning that I hope to illuminate. There is also much material under the headings of neuroscience and psychology that impinges on the theme and that I have ignored or mentioned only cursorily. For this my excuse is that these findings, sound and hard-won though they are, leave very major questions unanswered, and unravelling of the working of the brain will require combined assaults from many directions, of which I hope that what I have presented counts as one.

Reading, UK, 2008   Alex Andrew

Contents

Chapter 1
Cybernetics: Origins and Aims

Origins

The subject of cybernetics was announced by that name with the publication of Norbert Wiener's book (Wiener 1948). Of course, there were earlier stirrings including the influential paper of McCulloch and Pitts (1943) and other activity much earlier. A great deal stemmed from Warren McCulloch's lifelong quest that he epitomised (McCulloch 1960) as the attempt to answer the question: "What is a number, that a man may know it, and a man, that he may know a number?" Rather more concisely, his aim has been expressed as: "understanding man's understanding".

This can also be seen as the aim of much theorising that went before, under headings of philosophy and psychology, but McCulloch brought a fresh emphasis on experiment, particularly in neurophysiology. He also linked such studies to developments in technology, especially in automatic control and computing. The term "experimental epistemology" was coined, and it was recognised that an important aspect of the new approach was its interdisciplinary nature. This was sometimes illustrated by saying that the reflex arc of the neurophysiologist is essentially the same as the feedback loop of the control engineer, and this illustrates the general idea (even if closer examination shows the correspondence to be slightly tenuous). The interdisciplinary nature was acknowledged in the latter part of the title of Wiener's book where reference was made to "the animal and the machine".

It was also realised that the brain is a supreme example of a complex interconnected system that interacts with other complex systems, and consequently there has been much overlap between cybernetics and General Systems theory. The spanning of disciplines is now seen as not only linking neuroscience to physics and engineering but as also including such topics as sociology, management, and ecology.

Norbert Wiener was stimulated to consider these issues by his wartime work during the Second World War on predictors for antiaircraft gun control. Predictors that made a straight-line extrapolation of an aircraft's flight were standard, and Wiener's task was to examine the possibility of extrapolating a curved path as such. He then became impressed by the adaptability of pilots who could fly so as to avoid the presumed extrapolation. Wiener's findings were never in fact implemented in actual predictors, possibly because of the realisation that pilots would still outwit them.

A.M. Andrew, *A Missing Link in Cybernetics*, IFSR International Series on Systems Science and Engineering 26, DOI 10.1007/978-0-387-75164-1_1,
© Springer Science+Business Media, LLC 2009

(Since then the problem has been solved in different ways, first by the use of shells that would explode at a required height without necessarily hitting their target, and later by the invention of homing missiles.) Nevertheless the mathematical treatment produced by Wiener is a major contribution to the theory of automatic control, and its elaboration in the context of human interaction makes it germane to cybernetics.

At a more general level, as elegantly reviewed by Seymour Papert (1965), it can be argued that a range of technical developments introduced considerations that were not readily treated within the existing scientific disciplines, and cybernetics emerged to fill the gap. These considerations were associated with the manipulation and transmission of information, which was shown to be, at least according to one interpretation, measurable in units like a physical quantity (Shannon and Weaver 1949). There was a connection with physical entropy, but not in a way that immediately made sense to engineers concerned with heat engines. Feedback control depends on information made to flow in a particular way and is fundamental to servomechanisms and homing missiles and to a vast range of biological processes. Warren McCulloch's lifelong quest integrated nicely with the new discipline that these considerations prompted.

I am aware that I have given greater attention to the contribution of Warren McCulloch than to that of Norbert Wiener. It happens that I am more familiar with McCulloch's views and can comment on how they bear on what I now want to convey. For many years anyone wishing to discuss the topics at their source had unfortunately to join the camp of one or the other but not both of these founding fathers because there was a rift between them. The cause of the rift was not any disagreement over their scientific views, and at the time of launching of cybernetics they were firm friends and collaborators, as confirmed in the definitive biography of Wiener (Masani 1990). Masani acknowledges that the collaboration with McCulloch and Pitts was entirely positive for Wiener. The bias toward McCulloch here is not intended to be partisan, except insofar as it may compensate for the opposite bias in many other publications. There are of course many other people who contributed, as acknowledged by Wiener and McCulloch in their publications.

Early interest in cybernetics, and my own introduction, were with a strong emphasis on neurophysiology. The wartime requirements of radar were met by development of circuits to deal with pulses and other special waveforms, paving the way for electronic digital computers and also for recording equipment and stimulators applicable in neurophysiological research. Many bright young people had become acquainted with electronics, some of them after direction into this field when their first choice under normal circumstances would have been a biological subject. When the Second World War was over, the prospect of applying the new knowledge in the constructive and arguably humanitarian field of neurophysiology was extremely attractive. Means of recording from single nerve units, and of stimulating them, were developed, and it looked as though the nervous system could be analysed as though it was a computer or other electronic device, essentially as visualised by McCulloch. Since that time enormous advances have been made, but nevertheless much about the working of the brain and nervous system remains mysterious.

It should be mentioned that the word "cybernetics" (as "cybernetique") was used much earlier by Ampère and even earlier by Plato and Aristotle. Wiener claimed that he was not initially aware of the earlier use and my inclination is to accept this, but alternative views are reviewed in an appendix to this chapter.

Understanding

Warren McCulloch's aim has been denoted as "understanding man's understanding", and the term "understanding" needs clarification. The word has slightly different connotations in its two appearances: the first suggests a state of finality equivalent to "explanation" whereas the second suggests mental activity more generally, perhaps with a qualification that it should be to some extent rational. In either sense the term implies construction or acceptance of theories, whether recognised as such or referred to by some term such as "worldview".

As argued by Popper (1972) and more recently by others under the heading of "constructivism", a theory or worldview is necessarily an invention. In the context of the so-called exact sciences it is generally assumed that the current theory is the choice, of those devised to date, that best fits observations, perhaps fitting exactly within the precision of measurement. There is of course no guarantee that the theory will fit all future observations, and in any case there are other considerations that bear on the acceptability of a theory.

To say that something is understood or has been explained can only mean that a theory has been devised that seems, to a particular observer or group of observers (possibly including everyone who has given attention to the matter), to be highly satisfactory. Confidence is increased if the theory allows accurate predictions and if it provides a basis for successful manipulations that might include therapy in the case of brain theory.

However, in the case of Warren McCulloch's quest to understand man's understanding, no definite end point can be specified. The difficulty is increased by the recursive nature of the goal, epitomised by saying that a theory of the working of the brain would have the unique feature of being written by itself. This recursion must be a feature of neuroscience and Artificial Intelligence. (The use of capitals in referring to Artificial Intelligence, or AI, is to denote the subject area as now understood rather then the direct implication of the words.)

Although an end point to McCulloch's quest has to remain vague, there can probably be general agreement about progress towards it, and a great deal has been achieved. I am hoping, in this book, to show how some of the effort has been misdirected and so possibly to help set a course for the future.

Theories

The meaning of "understanding" depends strongly on the view taken of scientific theories, and it is worthwhile to make a digression to consider this in some detail, if only to clarify my own underlying assumptions.

As pointed out by Popper (1972), a theory can never be proved correct. If a perfect theory were to be formed, there would be no way of recognising it as such.

Popper goes further and denies that there is a formalisable deductive method by which theories can be derived from observations. He accepts that although a theory can never be verified it may be falsified, and a theory is only accepted as scientific or empirical if it is in principle falsifiable. He also shows, however, that falsification is not so simple as it appears at first sight, since an aberrant theory can be modified or supplemented so as to remain acceptable. An early example would be the survival of Newton's theories of gravitation and of motion following the observation that a feather falls more slowly than does a brick, where the theories were saved by bringing in the additional consideration of air resistance.

Especially when investigating an existing complex system such as the nervous system, there is no feasible alternative to forming, and then testing, quite bold theories. It is easy to have misgivings about such a method, and some wild flights of fancy have rightly been criticised as producing "theories looking for facts to fit them". However, Popper's claim that there is no formalisable method of inductive inference supports the view that progress has to depend on exercise of imagination.

A further fundamental uncertainty, referred to by Masani (1992) and others, stems from the fact that all science depends on the unprovable assumption that the future will have characteristics in common with the past. All forms of adaptation and learning can be regarded as implicitly embodying this assumption. Masani (2001) also claims that scientific thought has more in common with religious faith than is generally supposed. Though this view is unlikely to appeal to scientists, it is undeniable that the "experimental epistemology" approach depends on acceptance that the working of the brain can be explained in mechanistic terms without participation of some "vital force" unique to living systems. My own inclination is to go along with this, but with admission that it has the character of an act of faith.

Another criterion by which theories are judged is conciseness, where the choice of a short formulation is sometimes referred to as application of Occam's razor, the reference being to the philosopher William of Ockham who was born in Ockham in Surrey, England, and lived from c. 1280 to c. 1348. A theory is necessarily a succinct representation of a mass of sensory data, and it is reasonable to regard brevity as a virtue. A succinct theory is more likely to allow extrapolation to situations not previously encountered and thus predictions that inspire confidence if they turn out to be correct. Sometimes the razor principle is advanced as an assertion that nature is basically parsimonious and that a short theory is intrinsically preferable to a longer one, but this cannot be more than speculation. The matter of conciseness is discussed by Gaines (1977), with a review of relevant philosophical viewpoints. Gaines refers to an argument between advocates of respective models, of whom one may claim "My model is a better approximation", but the other could counter with the claim that his is simpler, and thereby further from being essentially just a record of the observed behaviour.

Rosenbrock (1990) has referred to the fact that many physical laws have alternative versions of which one is essentially causal and the other variational, and he extends the idea to biological and social systems. Well-known variational principles in physics are those of Hamilton, usually applied in dynamics, and of Huygens in optics. There can be no fundamental superiority of a variational principle over the causal alternative typified by Newton's laws, because the two versions are equivalent in the sense that each can be derived mathematically from the other. For a particular purpose one version may be more convenient than the other, but there is also something in human nature or culture that encourages preference for the causal alternative. This is a further way in which acceptance of a particular formulation is somewhat arbitrary.

In recent years, the realisation that a theory or worldview is essentially an invention of the observer has been discussed under the related headings of second-order cybernetics of and constructivism, the former being defined as "cybernetics that includes the observer" and the latter as the view that an observer's worldview is his or her own construction. Second-order cybernetics and its special flavour of constructivism originated with Heinz von Foerster (von Foerster and Poerksen 2002; Poerksen 2004), although he deprecated the "-ism" description. Von Foerster saw attention to the observer as closing a loop that had previously been ignored.

Some discussions from a constructivist viewpoint actually seem to deny, or at least to de-emphasise, any contribution from anything external to the observer, a view vigorously contested by Masani (1992, 2001). It is impossible to see such "radical constructivism" providing a basis for everyday functioning and indeed survival, and a more reasonable view is that each construction is formed subject to constraints imposed by the environment. This is consistent with Popper's assertion that a theory (or a worldview) can never be verified but should be in principle falsifiable.

One implication of these observations on theories is that there is no way of getting at anything that can be called absolute truth. On the other hand, access to absolute truth is not necessary in order to operate effectively and to feel satisfaction with explanations. We do have some knowledge of the external reality that must be assumed to exist, expressed as its conformity to a particular theory over the set of observations to date, even if the criterion of this varies from person to person. The access to reality is indirect and the tentative nature of theories must be kept in mind, but the achievements of science are undeniable and really do tell us something about the universe (Andrew 1993, 1994a, 1994b).

Neuroscience

Warren McCulloch considered "man's understanding" from many points of view. He was well acquainted with relevant philosophical works and at one stage began to train as a minister of religion. Later, as a psychiatrist he analysed individual behaviour. However, his emphasis on "experimental epistemology" found expression mainly in examination of the working of the brain and the rest of the nervous system at the level of individual neurons and neural connections. This can be seen from preparations he made in what he intended to become an international research centre in Orange, New Jersey (see later), as well as from his important contributions to neuroanatomy using the technique of strychnine neuronography, and from projects in the group around him, first in Chicago and then at the Massachusetts Institute of Technology.

There is a fundamental assumption that the brain is the seat of intellect. A great deal of evidence supports the view, including the effects of injury to the brain and recent observations using scanning techniques, though it is possible that other parts of the body are also involved. Reference is often made to a "gut feeling" and there is certainly some intimate connection between mental processes and the viscera, as well as to the endocrine system, but the brain seems to be the main seat of processing. Attention has been drawn to an important part played by the reticular formation of the brain stem (Kilmer, McCulloch, and Blum 1968, 1969), which suggests that this structure should be included, and the convenient term "brain" should, strictly, be replaced by "central nervous system".

Strychnine neuronography is a technique that allows mapping of pathways within the nervous system. An application of strychnine at a particular point evokes propagated "strychnine spikes" that reveal connecting paths. The spikes are conveyed antidromically, or in the opposite direction to that of normal transmission, and so do not usually cross synapses from one neuron to another. This is advantageous in allowing stage-by-stage mapping. The importance of McCulloch's early contributions is confirmed by an acknowledgement in the preface to the third edition of a classic text on neurophysiology (Fulton 1949) in its review of significant advances since the previous edition.

Many other techniques have been brought to bear on examination of the nervous system, many of them depending on electrical recording of neural activity, either noninvasively by surface electrodes (electroencephalography) or by electrodes placed close to a neural structure or even penetrating an active neuron. The necessary microelectrodes are made according to a variety of methods, the smallest being micropipettes of glass drawn to less than a micrometre diameter and filled with a conducting solution. Others to record from single units without penetration can be relatively crude, with metal cores. They can also be used for localised excitation by a pulse of current, and micropipettes have been used both for this and to inject minute amounts of transmitter substances that play a part in transmission across synapses. The importance of chemical specificity of synapses is an aspect that was not recognised in early work.

Some of the early work tends to be overshadowed by findings from tomographic scanning of the brain, of which positron emission tomography (PET) and nuclear magnetic resonance (NMR) scanning are the best known methods. These allow fairly precise location of neural activity in the brain and have the great advantage of being noninvasive (except for an initial injection in the case of PET) so that observations can be made on conscious subjects exposed to stimuli and performing mental tasks. A great deal of valuable and surprising data has come from such experiments, though as the methods operate by detecting metabolic changes associated with nervous activity, they cannot be expected to reveal the full picture at the single-neuron level. It follows that, valuable though these methods are, they are complementary to the earlier methods of investigation rather than supplanting them.

Visual Pathways

The visual pathways have proved to be particularly amenable to electrophysiological investigation. Much early work was done on the visual systems of frogs and other amphibia. A study that aroused special interest was by members of the McCulloch group (Lettvin *et al.* 1959, 1961) that related responses of units in the frog retina to the animal's survival needs. (The retina, in frogs and humans, is not just an array of photoreceptors but also has layers that process the image prior to transmission by the optic nerve.) These findings may have encouraged Hubel and Wiesel (1962, 1963, 1968) at nearby Harvard to record from the visual cortex of the brains of mammals while applying visual stimulation. Their work has been the basis of an enormous amount of further study that has revealed mechanisms of analysis of visual images and has indicated the extent to which their development depends on visual experience. This is an example of the "nature versus nurture" question that arises repeatedly in biological studies and has been studied particularly by Blakemore (Blakemore and Cooper 1970; Blakemore 1977).

The visual cortex has been found, in these experiments, to have four main types of units, termed respectively concentric, simple, complex, and hypercomplex. The area in the visual field to which a concentric unit is sensitive is, as the name suggests, centred on a point, with different sensitivity at different radii. The so-called simple units apparently combine the responses of a set of concentric units so as to respond to appearance of an edge, or a slit (a band lighter than its background), or bar (a band darker than its background) in the field. The so-called complex units also respond to edges, slits, or bars when they move, and hypercomplex units have the additional requirement that the edge, slit, or bar be of limited extent. All except the concentric units therefore have orientation in the visual field, and it has been found that, penetrating through the depth of the cortex (about 2 mm) at any one point, the units encountered have similar orientation. There are said to be orientation "columns" forming a mosaic over the surface. There is also a superimposed mosaic of ocular-dominance columns, characterised by a particular degree of dominance of one or other eye.

For the current purpose the details will be ignored, especially as they could fill several books, but it is clear that a great deal has been achieved. A further interesting aspect is that attention has been given to the problem of how the mosaic patterns (sometimes more expressively described as patterns of stripes) could have been generated (Miller 1998). This involves consideration of two-dimensional interactions in ways that can also account for other biological patterning including the stripes of the zebra. It is easy to accept that these studies show progress towards McCulloch's goal of "understanding man's understanding", but there is also much that remains mysterious. One worker epitomised the situation by remarking that he could explain the stripes of the zebra but was still puzzled by the horse part underneath.

Microtubule Computation

There are still very large gaps in our knowledge of physiology in general and of the nervous system in particular. It is even possible that the accepted "neural doctrine" may not be the basis of all neural processing. Cells of the body, including neurons, have within them microtubules. There is reason to think that these have a communication function, especially as they have directionality. It has been found that molecules are conveyed along them, some in one direction and some in the other, which supports the idea that they provide slow communication channels.

There are suggestions, however, that microtubules also convey rapid signals in a way that depends on quantum theory for its explanation (Dayhoff *et al.* 1993; Jibu and Kusio 1994; and other papers in Pribram 1993). Dayhoff *et al.* argue that microtubules provide a mechanism for backpropagation, a learning principle that has been widely used in applications of artificial neural nets (Rumelhart *et al.* 1986; Andrew 1993b) and is discussed further in Chapter 5.

The treatments in terms of quantum theory are highly speculative, but support for the suggestion that microtubules play a major role comes from observations by Allison and Nunn (1968) and Allison *et al.* (1970) that substances used as inhalational anaesthetics have a strong but reversible effect on microtubules. Their experiments were on *Actinosphaerium nucleophilum*, which are protozoans found in pond water.

Sejnowski (1986) has shown that complex neural computations are carried out in a time that allows only a remarkably small neural "depth", assuming the usual synaptic delay of about a millisecond. Of course, this is not a problem if the computation is performed on a high-speed network of microtubules within neurons and perhaps other cells, rather than by the network of neurons as in classic theory. Despite the speculative nature of these theories, it is salutary to realise that our understanding is still at a stage where there can be debate.

Memory

Other aspects that are still mysterious are the mechanisms underlying memory, especially long-term memory. That short-term and long-term memory depend on different mechanisms is indicated by the differing effects on them of degenerative diseases and of such assaults as concussion or the electrically induced convulsions that were once much used as psychiatric therapy.

The nature of long-term memory storage is uncertain, despite the insight given by von Foerster (von Foerster 1950; von Foerster and Poerksen 2002). He made a case for supposing it to be molecular but there is much that is unexplained about how this interacts, in both directions, with short-term memory. Von Foerster's first contribution was to derive a law for rate of forgetting based on quantum theory, but this did not at first fit the experimental data obtained by psychologists. However, when the theory was modified to take account of periodic refreshment of the remembered items by recall, the fit was good. The possibilities of recall and refreshment imply two-way exchange between the two forms of memory.

It is not even clear whether long-term memory resides in the neurons or in other structures. Likely mechanisms for short-term and rather longer-term memory depend on neural reverberations and on changes in the synapses, but there has been speculation about other possible storage mechanisms, separate from the neural circuitry (Byrne 1970). The success of geneticists in accounting for storage and transfer of information by DNA has naturally encouraged speculation that something similar may underlie psychological memory. A molecular storage mechanism is an attractive idea because of the vast information capacity of the brain, though synaptic modifications throughout the unimaginable number of neurons in the brain might possibly also account for it. Some experiments have produced evidence of chemical transfer of learning between individuals, but the findings are disputed. The main results were obtained on planarian worms, though claims have also been made for such transfer in rats. It is clear there is much still to be unravelled about the physical nature of memory.

C fibres

Of the many unanswered questions concerning the brain and nervous system, just one more will be mentioned. It is possibly a rather arbitrary choice but has a connection with important recent developments. The nervous systems of humans and other animals contain large numbers of very fine fibres referred to as C fibres. No function was ascribed to them until a suggestion was made by Patrick Wall and colleagues (Wall, Fitzgerald, and Woolf 1982; Melzack and Wall 1988). This was in connection with an earlier demonstration (Wall and Egger 1971) of neural plasticity, demonstrated by them for the first time in an adult mammal.

The term "neural plasticity" is applied to any long-lasting change in the properties of neurons as a result of their history. Such effects are of course interesting in connection with theories of learning and self-organisation, though Wall was careful to avoid rash claims about such significance for his findings.

Wall and Egger worked on rats and recorded from an area in the lower part of the brain on which the body surface is mapped. The area is only a few millimetres across, but with microelectrodes they were able to trace out, on each side, mappings of the fore and hind limbs on the opposite side of the animal. For a particular position of the microelectrode, the surface of the animal's body was explored with moderate pressure until a point was found where pressure produced a response.

In a number of animals these workers interrupted, on one side, the nervous pathway connecting the hind limb to the central nervous system. If the map was examined within a day of the operation, there was a silent area where the limb would have been represented. If a longer time was allowed to elapse (from one to three weeks for the full effect), the representation of the forelimb was found to have expanded, as though drawn on a rubber sheet and stretched, into the area formerly occupied by the representation of the hind limb.

One way in which this might have occurred would be by physical shrinking of the idle neurons and migration of others into their space, but reasons are given for believing that the effect was due to a rearrangement of connections to neurons remaining in place. There is then a choice between the sprouting of new connections and the activation of previously existing dormant ones, and reasons are given for preferring the latter view.

At this stage, it looked as though the "rubber-sheet" effect was due to cessation of activity in the neurons that previously mapped the disconnected hind limb, but further experiments were not consistent with this. It was found that the effect was not seen when the pathway was interrupted farther back, even though this produced the same cessation of activity. The effect also depended on just how the interruption was effected, whether by cutting or crushing or freezing. The findings were consistent with the view that the "rubber-sheet" effect depends on interruption of the numerous C fibres that accompany the larger ones carrying the main sensory signals. The C fibres apparently play some role in organising the connections of the main fibres. Supporting evidence came from experiments using capsaicin, the substance responsible for the pungent taste of hot peppers and chilis, which selectively destroys C fibres.

This view of the role of C fibres is controversial and in any case does not spell out their role in detail. This state of affairs further underlines the degree to which the nervous system is still mysterious.

Orange, New Jersey

Warren McCulloch's faith that electrophysiology would help his quest found expression well before the time at which cybernetics emerged by name and well before the collaboration with Walter Pitts that gave rise to the 1943 paper. I know this because of a unique and privileged experience at a date in the mid-1950s, when I accompanied Warren on a visit to Orange, New Jersey. The purpose was to hand over a large house to a new owner. The house belonged to Warren and his wife Rook, and much of it had been built by them with the help of neighbours. It was on a magnificent scale and included a hall for lectures or concerts having the height of two storeys. At the back of the stage there was an enormous window with a view through forest down to a lake.

The house had been built when both Warren and Rook expected to inherit wealth that would allow them to run their own research centre to which they would invite, at their expense, visitors from anywhere in the world. However, the finances of both families suffered in the depression of the 1930s, so that the scheme had to be abandoned and Warren had to find paid employment. They did not part with the house immediately, but the visit to hand it over was the sad acceptance of the end of a dream.

There was space under the pitched roof that had been intended to house laboratories. It was expected, at the time of building, that the experiments would involve recording bioelectric events using mirror galvanometers. To provide steady supports for these, in a house of timber-frame construction, each chimney-stack had a stout brick shelf protruding from it. This was an intriguing indication of interest in bioelectric phenomena prior to invention of the cathode-ray tube.

McCulloch's confidence that neural events could explain mental performance was strengthened by the later results (McCulloch and Pitts 1943) showing that model neurons could be the basis of any finite automaton. The result has also been demonstrated more transparently by Minsky (1967). It will be argued here, though, that it does not follow that all brain function can be accounted for in such terms, and that the argument depends on undue acceptance of "logic", as usually understood, as the basis of thought.

Artificial Intelligence

One way of demonstrating progress toward McCulloch's goal of understanding man's understanding is to construct machines that model human mental ability. That is to say, a main aim of cybernetics should be to achieve artificial intelligence (George 1977). At first it was widely assumed that appropriate machines would have the form of nets of model neurons, either as specially constructed hardware or as simulation programs in general-purpose digital computers. A great deal of effort was put into this in the early days of cybernetics, with its most tangible result the perceptron principle pioneered by Rosenblatt (1961) and treated very clearly by Nilsson (1971) among many others.

It was soon believed, however, and strongly argued by Minsky and Papert (1969) that this approach was of limited value, though in recent decades there has been a very strong revival of interest. The topic of Artificial Intelligence was introduced by that name at a conference in 1956 and represents the attempt to achieve intelligent performance by artifacts without restriction to a neural net or other particular architecture.

Artificial Intelligence, or AI, is shown with capitals to indicate the new connotation, which has come to be associated with particular programming and knowledge-manipulating techniques rather than with simulation of brain mechanisms in detail. Important developments in computer science have been stimulated by the needs of AI. This is hardly surprising as the topic could be defined as the search for ways of making machines do whatever brains can be seen to do but that is currently difficult for machines. The subject has an intrinsic fascination that has attracted many able people to work in it, and it has been described as: "the department of clever tricks of computer science".

The achievements of AI are not to be dismissed lightly. One of the early contributions was to emphasise the importance of heuristics, or inputs in the nature of hints or suggestions. Problem-solving methods depending on heuristics are sometimes introduced as "methods that do not always work", thereby differing from algorithmic methods. This distinction is not always quite correct as heuristic methods may be used to guide a search for solutions in a situation where a solution will eventually be found irrespective of the heuristics, but with the search time strongly dependent on heuristic guidance, perhaps so much so that unguided search would require a prohibitive amount of time.

The AI effort has given new feelings for the amount of difficulty involved in particular tasks. One impressive example, among a vast number, is new appreciation of the difficulty of analysing natural language text, where semantic and syntactic considerations must interact, and interpretation is likely to depend on a wide background of knowledge, including assumptions about the speaker's or writer's intentions. People perform such analysis without being aware of most of the details, but they are brought to light by the attempt to automate the process. There is a stark contrast with the impression of a simple regular structure given by the standard school treatment of grammar. Similar considerations apply to visual scene analysis, and to the analysis of speech or handwriting, of which the latter two must operate in conjunction with the preciously mentioned analysis of text.

One of the goals of the AI effort has been to make powerful chess-playing programs, and one has now been written that challenges the best of human chess masters. This was achieved after a considerably longer time than was predicted by enthusiasts in the early days of AI, but the milestone has been reached. The operation of the computer program is cerrtainly significantly different from the human mental processes applied to the same task, though there may be enough in common that something can be deduced about human processing.

On the other hand, there are many things that people do very easily that are beyond the current capabilities of machines. Some of these become apparent in considering things we might like domestic robots to do, like tidying and cleaning after us at home (but tidying with discretion, so that we still find things), or weeding the garden or running an errand by travelling safely on the street and making a required purchase. Marvin Minsky, for many years the foremost figure in AI, has even described the subject as brain-dead (Minsky 2003) though with the caveat that a current project of his own and another of Douglas Lenat could revive it.

There is, certainly, reason to believe that AI has somehow failed to fulfil expectations and does not give the hoped-for vindication of cybernetics. It will be argued in Chapters 2 and 3 that this is at least partly attributable to a wrong starting point. There are many ways in which the nervous system processes continuous variables, and such processing can be seen as more primitive than the concept-based kind. As evolutionary developments do not erase all traces of previous forms, the mechanism underlying concept-based processing must have features inherited from continuous processing and will not be successfully modelled without recognition of this.

Early work in AI focused on such discrete problems as playing of board games and theorem-proving, the latter much influenced by the work of Polya (1954) on induction and analogy in mathematics. These are clearly forms of human mental activity that are well removed from the conditions under which intelligence evolved, and it will be argued that the limited success of the approach can be at least partly attributed to this lack of attention to what might be called paleo-AI.

The distinction between continuous and discrete processing might also be related to the much-discussed philosophical problem of the relationship between mind and body. Continuous processing is a feature of the autonomous or vegetative regulatory functions that operate independently of consciousness and are associated with "body", whereas discrete or concept-based processing is part of what is understood by "mind". It is, however, only a part, and the argument here is that interaction between the two kinds of processing is more significant than is generally assumed.

The emphasis of McCulloch and Pitts on discrete logic can be seen as similarly inappropriate, as can also the focus on concepts in Pask's Conversation Theory (Pask 1975). These aspects will be treated at length in Chapters 2 and 3.

Wider Applications

Apart from neuroscience and AI, which have a clear connection with McCulloch's stated aim, there are other important developments under the heading of cybernetics that do not follow quite so directly. The brain is a supreme example of a complex interconnected system and as already mentioned the rise of cybernetics has been associated with that of General Systems theory, as well as being part of a general acceptance of the value of an interdisciplinary approach that includes sociology, management, and ecology.

A major extension of the area of interest was due to Stafford Beer (1959, 1962, 1975, 2004) who applied principles of cybernetics to management. (The 2004 paper reviews ten major works by Beer and is in a journal issue that contains other papers as a posthumous tribute.) In the 1962 paper he drew parallels between the organization of the central nervous system and that of a firm. There are significant correspondences between the two, particularly in the sharing of action and responsibility between top management and distributed decision-makers as a means of coping with complexity. The nervous system has the brain in overall control of at least certain functions but with a host of spinal and other reflexes that can operate independently and in fact become more powerful when disconnected from the brain.

The relationship between the brain and these reflexes is complex, and under some circumstances the brain can overrule or modify a reflex (Melzack and Wall 1988; Andrew 1995b). This is illustrated by the fact that a person who picks up an expensive coffee cup that proves to be unexpectedly hot is usually able to avoid dropping it immediately. The reflex arc does not depend on the brain, but the brain plays a permissive role. There is an obvious correspondence to the delegation of responsibility in a business, where rapid responses are needed despite complexity. In numerous publications, and particularly his 1975 book, Stafford Beer has referred to "complexity engineering" in a wide range of practical contexts. This is related to Ashby's principle of requisite variety (Ashby 1960), which essentially states that a complex situation demands complex control responses.

It is, however, apparent that the analogy between business and social systems on the one hand and neural or other biological models (such as an ant colony) on the other are only useful up to a point. For one thing, these systems do not accord to their constituents the rights that are believed to be appropriate to humans. No one, for example, wants to be treated like a honeypot ant, and in many other biological systems the constituents are treated as expendable. Biologists even have a term, "apoptosis", to denote the programmed death of a biological cell, as a feature of multicelled organisms including humans. (The term "apoptosis" is chosen to mean "falling off" as leaves do in autumn, this being one example of programmed death.)

Apart from this, though, people operating in business or social systems have an awareness of structure that is pretty certainly not shared by neurons. According to Stafford Beer, self-organisation of a business is determined by "algedonic feedback" or feedback associated with pain or pleasure. That is to say, if an action produces a bad result, equivalent to pain, it should be avoided in future, and an action that produces a favourable result should be repeated. This might seem an obvious policy for a manager, but its implementation requires some appreciation of structure of the situation. There is an implicit assumption that the manager has some feeling for what actions are relevant to performance of the enterprise, otherwise he could waste time and effort by looking for connections between the feedback and every trivial action or choice he made, like whether to have sugar or sweetener in his coffee. Also, any effect of an action is only apparent after a delay and may be spread over a substantial time interval. It is only by knowing something of the structure of the system that the effect can be associated with its probable cause and the feedback loop closed.

It is usual to assume that neural nets organize themselves without the constituent elements having the benefit of this appreciation of structure. (Certainly, it is dangerous to make assumptions about neurons, especially assumptions of simplicity, but if they do have this awareness of structure there must presumably be a lower level at which its emergence has to be accounted for.) In theories about neural adaptation, a criterion of goal-achievement has been postulated as "affective tone" by Wiener (1948), and the attractive term "hedony" was coined by Olive Selfridges. It is essentially the same as Beer's algedonic feedback, but it is acknowledged in treatments of neural nets that the operation of its feedback has to be a more complicated matter.

Beer's "viable system model" has found wide and valuable application in management. One of its features is an emphasis on recursion, such that a viable system contains nested viable systems and may itself be nested in a supersystem. People, with their nervous systems, are viable systems nested within a social or business system, and their neurons and other cells are viable systems nested within them.

There is, however, a drastic change in the nature of the viable system at a particular stage of nesting, namely in going from a system with human participants to the nervous system. An undue focus of attention on systems with human participants misses important aspects in much the same way as they are missed in the undue emphasis on discrete logic in other studies, including those of McCulloch and Pitts and of the originators of Artificial Intelligence. In the next two chapters an alternative starting point will be defended, requiring a focus on continuous processing.

Appendix to Chapter 1: Early History

The term "cybernetique", equivalent to "cybernetics", was used by the physicist Ampère in the early 19th century to denote the science of politics in a classification of human knowledge. Similar use of an equivalent term, though with a slightly different emphasis, was made by the Polish writer B.F. Trentowski in 1843.

The original Greek version of the word, meaning "steersman", was also extended in meaning to indicate political or social control by both Plato and Aristotle. These pre-Wiener uses of the term are reviewed in a scholarly paper by Mihram *et al.* (1978) who argue that Wiener should not be credited with originating cybernetics, and also that the subject should be seen as primarily concerned with politics rather than with computers and neurophysiology. Wiener's claim that he was originally unaware of the earlier uses is viewed with surprise and even a degree of scepticism in view of his undoubted erudition. The use of the term by the ancient Greeks is also discussed at length by Johnson (2008).

Despite these scholarly arguments, I have persisted in accepting Wiener's assertion that he invented the term independently, and that the new version of cybernetics was stimulated by his wartime work on antiaircraft predictors and by developments in analog and digital computing and neurophysiology, with his attention to the last-mentioned stimulated by McCulloch and Pitts. According to this viewpoint, computing and neurophysiology are of primary concern, though the interdisciplinary nature of the subject and its wider applicability to, for example, social studies were recognised by McCulloch and Wiener from the start of the reincarnation and reorientation launched by them.

Summary of Chapter 1

Aims and origins of cybernetics are reviewed, with particular reference to Warren McCulloch's declared lifetime quest of "understanding man's understanding". The meaning to be ascribed to "understanding", especially in this self-referential context, is acknowledged to be elusive, such that McCulloch's quest can have no well-defined end point. Nevertheless there can be general agreement about what constitutes progress towards it achievement, and important contributions to neuroscience by McCulloch and his followers, termed by them "experimental epistemology", are reviewed, as well as associated developments in Artificial Intelligence. It is shown that there is much about the brain that is still highly mysterious. Some failures of understanding can be ascribed to undue readiness to accept discrete logic as fundamental.

Other important developments in cybernetics do not follow so directly from McCulloch's stated aim. The brain is a supreme example of a complex interconnected system and the rise of cybernetics has been associated with that of General Systems theory, as well as being part of a general acceptance of the value of an interdisciplinary approach. The spanning of disciplines is now seen as not only linking neuroscience to physics and engineering but as also including such topics as sociology, management, and ecology.

It is, however, argued that some arguments that have been used in linking the topics are insecure. In particular, the mapping of nervous control onto a management system is only valid up to a point and can be misleading. The people in an organisation have awareness of structure in a way that is not paralleled at the neural level. Ignoring this has something in common with the acceptance of discrete logic as fundamental, and an alternative approach will be developed in later chapters.

Chapter 2
Where to Start?

Brains and Computers

It is certainly not obvious where to start in a description or study of "understanding" or "intelligence". The terms are roughly synonymous, the main difference being that the second carries rather more suggestion of action. One feature that the brain has in common with a modern computer is a degree of versatility such that no short description can be given of the capabilities of either and no clear indication of a starting point for investigation or study. For the computer, one approach would be to go back to binary information storage and processing, and the use of stored programs, and to follow developments from there, roughly in historical order. Actually, these elementary aspects of computing are now submerged under such a mass of later developments that they would probably not now be the chosen starting points for a textbook or a taught course, but at least we have access to the ideas and intentions of the pioneers of the subject, and examination of these offers one possible starting point.

For the brain there is no corresponding access to ideas or intentions of a designer, but following Ashby (1960) it can be seen as a product of biological evolution and hence of selection according to survival value, and this gives some indication of where we might start.

Discrete Logic

It is often assumed that discrete logic is the basis of effective thought, but it must have been a relatively late evolutionary development. The means of processing sensory data and using it to determine actions favouring survival is far from being peculiar to humans. The capability can even be ascribed to single-cell organisms, so is not dependent on a nervous system. Reactions of a particular single-cell organism are described by Alberts *et al.* (1989) and will be discussed in Chapter 4. The observation that microorganisms can show such behaviour gives indirect support to the suggestion, referred to in the past chapter, that microtubules may be involved in computations, possibly even when coexisting with nervous systems. This is of

A.M. Andrew, *A Missing Link in Cybernetics*, IFSR International Series on Systems Science and Engineering 26, DOI 10.1007/978-0-387-75164-1_2,
© Springer Science+Business Media, LLC 2009

course highly speculative, and it is impossible to know whether human intelligence is usefully considered as an evolutionary development from that of microorganisms.

Nevertheless, although the details are obscure, there can be little doubt that human intelligence evolved from something more primitive. The evolutionary viewpoint is important because new developments do not obliterate traces of previous forms, and traces of the primitive modes are likely to be effective "behind the scenes" in the intelligence of humans and other animals. There are many indications that human thinking is far from being fully represented by deductive logic, or in natural language, even if the outcome is often promptly rationalised and expressed linguistically. What can be expressed linguistically is just the tip of the iceberg of the thought process. The view is supported by no less a thinker than Albert Einstein in a letter to Hadamard quoted by Penrose (1989):

> The words of the language, as they are written or spoken, do not seem to play any role in my mechanism of thought. The psychical entities which seem to serve as elements of thought are certain signs and more or less clear images which can be "voluntarily" reproduced and combined . . . The above mentioned elements are, in my case, of visual and some muscular type. Conventional words or other signs have to be sought for laboriously only in a second stage, when the mentioned associative play is sufficiently established and can be reproduced at will.

Some elusive aspects of the thought process have been acknowledged by the emphasis on heuristics in mainstream Artificial Intelligence. The limited success of AI suggests that there are further aspects still more elusive, probably also describable as heuristics but of a kind not yet explored, and it will be argued here that attention to continuous processing may be a key.

An additional reason for rejecting discrete logic as the basis of effective thought is that it is only useful once some concepts have been formed. Attention has been given to this under the heading of "binding" in neurophysiology, but something more is needed. Here and in the next chapter it will be argued that the manipulation of continuous signals is an important part of neural processing and should be regarded as more primitive than concept-based processing. It is useful to consider how the use of concepts could have evolved from continuous processing, as an evolutionary development that can throw light on processes operating "behind the scenes".

Meaning of "Logic"

The word "logic" is usually assumed to imply the manipulation of discrete concepts and as such is the basis of much work in AI, and it is definitely understood in this sense in computer technology. However, the word also denotes the science of thought, or thinking about thinking (with, as one logician put it: "no essential difference between the thinking that is thought about and the thinking that thinks about it").

Much confusion has resulted from the common assumption that these two meanings of "logic" are equivalent. If they were, there would be no need to look for anything "behind the scenes" in the thought process. It is perhaps hardly surprising that humans tend to emphasize the discrete or concept-based aspects of thought, as it seems to be their superior capacity for it (or at least for processing that finds expression in such terms) that distinguishes us from other animals and has allowed the development of science, technology, philosophy, and so on.

Formal logic, like mathematics, is often preferred to natural language because of its apparent precision, but this is illusory if it depends on distorting the representation of a problem or situation to make it fit the formalism. The preference is reminiscent of certain advice given to young men in Victorian times, where they were urged to remember that white is white and not "a kind of light colour". Unfortunately for such a rule, colours described loosely as "white" vary considerably among themselves, as becomes clear when buying paint to match some existing decoration, and the less precise "kind of light colour" may be a more accurate description.

There is evidence, from Jerry Lettvin quoted by Anderson (1996), that Walter Pitts was unduly impressed by the possibilities of formal logic. It also seems clear from numerous papers in the earlier collection of the works of McCulloch (1965) that he attached much significance to the fact that the model neurons of the 1943 paper could perform the operations of formal logic. This is a somewhat tricky issue, as the operations of formal logic can form the basis of a general-purpose digital computer and can therefore underlie digital arithmetic and hence processing of continuous variables to any required accuracy. However, no evidence has been found of the nervous system using a binary (or decimal, etc.) system, and it is not useful to think about its processing of continuous data in such terms.

The last observation should be qualified by noting that it is possible to form a number system with radix one, and therefore no "carries", such that magnitudes are represented by the length of the number. This can correspond to the number of pulses in a burst, or the number of active elements in a bundle of parallel binary channels, and this can be seen as a meeting point of analog and digital representations. The analog interpretation is in a sense more "natural", however, and likely to be relevant to the evolution of intelligence. There is a connection with the emphasis on the "fabric" of computing by Beer (1959), which will be referred to again in the next chapter.

When McCulloch and Pitts (1943) produced the famous theory, they were well aware that their model neurons were unrealistic in many respects. One is that real neurons usually fire repetitively in a way that suggests the significant feature is the frequency and hence a continuous signal. It was argued, very reasonably, that it would be instructive to see what could be achieved with these highly simplified models of neurons, but reservations about the correspondence to real neural nets have often been ignored. Even the originators have arguably fallen somewhat short of the ideal in this respect.

There is much other evidence of an unwarranted prejudice in favour of discontinuous models of brain functioning, some of it reviewed in the next chapter.

Laws of Form

The making of a distinction, for instance between inside and outside a boundary, has been suggested as the most fundamental component of intelligence and is the starting point for the treatment elaboraterd in George Spencer Brown's *Laws of Form*

(Brown 1969). This is tantamount to advocacy of discrete logic so does not offer the most fundamental starting point. This is not to deny that it offers a valuable introduction to a large area of finite mathematics, but the value is to students who, along with the rest of humanity, have already progressed beyond the most elementary stage of development of intellect.

Associative Recall

A number of approaches appear to assume that the basic problem is achievement of associative recall. People certainly have a remarkable capability to recall remembered items using a wide variety of associations. Postulated mechanisms allowing such recall include neural nets as postulated by Hopfield (1982), and others depending on holography (Gabor 1968; Willshaw 1981), as well as on a method mathematically equivalent to optical holography but not requiring mechanical rigidity (Willshaw *et al.* 1969).

These theories require a prior mechanism to select the items that are worth remembering, so even if valuable in themselves they do not provide a starting point. Another objection is that the kind of similarity that is effective in forming their associations is inherent in the method and not modified by learning. The theories therefore fail to account for the subtleties of human learning, which operates on more than one level by allowing different criteria of similarity to be learned. That different criteria are effective in different contexts was demonstrated by reference to a simple task of pattern recognition by Selfridge and Neisser (1963), who showed that spatial correlation as a measure of similarity could be misleading in the classification of hand-blocked characters, even where the images were adjusted to standard size and orientation. Introspection readily supports the idea that different similarity criteria are used in different tasks. For example, those that are effective in classifying leaves of trees by their shape are not the same as are effective in classifying hand-blocked characters.

At the same time, as Pribram (1971) points out, and as discussed by Andrew (1997, 2000), the similarity (or invariance) criteria effective in elementary recognition of objects, irrespective, within limits, of orientation and size, can be argued to correspond closely to the achievements of holography. Though not central to the current search for a starting point for consideration of brain activity in general, it is interesting to note that there are alternative approaches to the study of visual perception, one with reference to holography and the other to analysis of the image in terms of constituent features. Pribram refers to experimental demonstrations of spatial frequency selectivity that support the holographic version, but it is also clear that he does not suggest one single holographic reconstruction as an adequate model. It is possible that the multilevel application he visualizes is reconcilable with a feature-based scheme.

The remarkable human facility for associative recall is also apparent in essentially verbal contexts, for example, where the comment "she didn't bat an eyelash" evokes a similar phrase referring to "eyebrow" but with the understanding that the substitution implies supreme imperturbability. These aspects have a bearing on the principle of "heuristic connection" treated in the next chapter, but they do not provide the needed fundamental starting-point.

The Binding Problem

In studies of neurophysiology, the way that sensory stimuli come to be associated, or bound together, as representations of recognised objects, has been termed the "binding problem" (Andrew 1995a). Experimental studies have failed to reveal special cells responding to discriminable categories, sometimes indicated by saying there is no "grandmother cell" (Gross 2002).

Experimental results bearing on the problem have been obtained by Singer (1990), who made simultaneous recordings from pairs of neural units whose outputs might be bound at a later stage of processing. The experiment was repeated with numerous variations, ranging from relatively simple cases where recording was from primitive cortical units as described by Hubel and Wiesel (1962), to others in which the paired units were in different brain hemispheres or responded to different sensory modalities.

It was found that, throughout this wide range of situations, the characteristic feature of paired discharges that were suitable for subsequent binding was a high degree of temporal coincidence of impulses in the two streams, indicated by a sharp peak at zero time in their cross-correlogram. The significance of this for underlying mechanisms is not easily judged.

Attempts to treat the binding problem theoretically have tended to focus on selection of patterns of activity that recur frequently. An early model of binding on this basis was made by Chapman (1959). A similar emphasis on recurrence appears in a comment by Einstein, quoted by Negoita (1993):

> I believe that the first step in the setting of a "real external world" is the formation of the concept of bodily objects and of bodily objects of various kinds. Out of the multitude of our sense experiences we take, mentally and arbitrarily, certain repeatedly occurring complexes of sense impressions (partly in conjunction with sense impressions which are interpreted as signs for sense experiences of others), and we correlate them to a concept – the concept of the bodily object. Considered logically this concept is not identical with the totality of sense impressions referred to, but is a free creation of the human (or animal) mind.

In considering origins of intelligence, presumably governed by needs of survival, it is necessary to look for something other than frequency of occurrence of patterns or complexes that leads to their "binding". At a later stage of evolution it is possible that patterns are chosen arbitrarily, as suggested by Einstein, and later retained only if they prove useful, but such a process (which could be described as operation of the heuristic of curiosity) cannot account for the emergence of binding and conceptual thinking in the first place. A view will be presented here that is plausible from an evolutionary viewpoint, consistent with Ashby's insistence that the brain should be seen as a specialised organ of survival.

Models

Craik (1942) viewed the nervous system as a calculating machine capable of modelling or paralleling external events. The idea of a model is not so simple as at first appears, because clearly a mental model is not a scaled-down version of something external, in the same way as, say, a model train is a scaled-down version of a physical entity. A model that can exist in memory, whether human or computer, has a less tangible relation to its prototype. The model and prototype correspond in only some of their characteristics, and the fact that the model is useful means that these characteristics are appropriate for some purpose.

Intelligent entities are usually viewed as working towards goals, or acting purposefully. Any such entity that is a product of evolution can be interpreted as pursuing the goal of survival and usually as pursuing other goals subsidiary to this. The way of interacting with the environment may be described as a set of rules, or a policy, and may or may not invite description as involving a model. The acceptance that a model is only justified in terms of a purpose suggests there is no firm distinction between a purposive system that forms a model and one that does not. In fact, any set of rules that favours achievement of a goal implies modelling of the environment because it amounts to the assertion that the environment is such that action according to these rules is likely to have a good outcome. There is correspondence here with the view of scientific theory expressed in the past chapter and with the general observation that access to absolute truth is not necessary in order to operate effectively and to believe that explanations are satisfactory.

Highly intelligent systems (of which we usually take ourselves examples) often form something that compellingly invites description as an internal model of a part of the environment. A person can sit still and imagine himself engaged in some activity, such as climbing a mountain or changing a wheel on his car or asking for a loan from his bank manager. If he is familiar with the environments in which these tasks are carried out, his mental images may contain quite a lot of detail. Even here, though, the model is related to purpose, and details that are not relevant to any purpose are usually absent from the image. It seems safe to assume that, in considering the early stages of evolution of intelligence, the focus should be on useful interaction with the environment rather than on modelling.

Of artificial systems for adaptive control, some are classified as forming a model and some not. The distinction is based on how a designer went about constructing the system and on how it is most readily understood. The overall performance of a model-based system may be indistinguishable from that of a system not explicitly embodying a model. This will be discussed in more detail in Chapter 4.

The observation that all adaptive control systems can be described as forming models helps to provide the "missing link" referred to in the title of this book. A servo system has more in common with the kind of cognitive system that is studied under the heading of AI than is generally thought. To achieve rapid and accurate response, a servo usually embodies more than simple negative feedback. A fairly usual expedient is to combine components denoted by the acronym PID, for "proportional, integral, and derivative", where straightforward negative feedback would have only the "proportional" component.

If all the properties of the servo and its environment (the load it will drive) are known, established theory allows the derivation of optimal gains for the three components (in the situation where everything is linear). Alternatively, the gains may be set by successive adjustment while the servo is in operation (possibly with additional components if there is serious nonlinearity). Humans and animals are remarkably good at making such adjustments, especially when they themselves constitute the servo. They do so in a wide range of manipulative tasks and learned skills, from walking to riding unicycles and flying jumbo jets. All of these involve modelling of the environment to a greater degree than is generally realized.

Conditioned Reflex

The phenomenon of conditioning, famously investigated by Pavlov, has been the basis of numerous psychological studies, as reviewed for example by Dickinson (1987). Pavlov studied in particular the occurrence of salivation in dogs, which is an unconditioned response to the appearance of food and is also produced as a conditioned response by another signal such as the sound of a bell when this has been repeatedly paired with presentation of food. Here the unconditioned response of salivation clearly has survival value because it helps the dog's digestion, and its elicitation by a different signal may be advantageous if the signal is indeed linked to food. It might for example allow salivation to begin sooner than if the animal waited for the materialisation of food as such.

This classic or Pavlovian form of conditioning agrees very well with the mechanism for synaptic modification proposed by Hebb (1949), expressed by him as follows:

> When an axon of cell A is near enough to excite a cell B and repeatedly or persistently takes part in firing it, some growth process or metabolic change takes place in one or both cells such that A's efficiency, as one of the cells firing B, is increased.

Another type of conditioning experiment, particularly associated with B.F. Skinner and his invention of the "Skinner box", is called operant conditioning. Here an animal is rewarded, typically by receiving a food pellet, when it performs a particular action such as pressing a lever. The first press is accidental as the animal explores its environment, but if it is hungry it repeats the action and receives more food. This offers a starting point for the study of intelligent behaviour favouring survival. An action of the animal is found to correlate with something favourable to survival and the response is to enhance the action. It seems reasonable to regard responsiveness to such correlation as a primary feature, perhaps the essential primary feature, of an intelligent entity.

The connection between "conditioned reflex" and statistical "conditional probability" was stressed by Albert Uttley (1956, 1959), who built a special-purpose computer to model the conditioned reflex by computing conditional probabilities

(Andrew 1959a). The conditional probability of a particular event, given that another has occurred, is an indication of the kind of linking referred to as "correlation", and in this context it is a more appropriate numerical indicator than is the correlation coefficient.

Uttley's conditional probability computer had a set of yes/no input channels, five in number as implemented, and arbitrarily labelled j, k, l, m, n. The device was "trained" by presentation of a succession of inputs consisting of nonempty subsets of the five channels (31 possibilities). It had 31 units that responded to respective subsets and counted the numbers of occurrences. The units counted all subsets, so that, for example, simultaneous inputs through the j, k, and l channels would activate the jkl unit and also those responding to occurrences of j, k, and l and of the pairs jk, jl, and kl. The counts in all seven units would be incremented.

Once sufficient counts had been accumulated, the device was able to make inferences of activity, based on computation of conditional probability as a ratio of counts. The conditional probability of activity in the l channel, given that there is activity in the j and k channels, is given by:

$$p_{jk}(l) = (\text{count in } jkl \text{ unit})/(\text{count in } jk \text{ unit})$$

and it was arranged that, when such a ratio had exceeded a preset threshold value, an inference would be made of activity in an input channel, here the one labelled l.

In the device as implemented the counting was done in analog fashion, by incrementing the charge on a capacitor in each unit, and it was arranged that the charging law was roughly logarithmic, so that the division needed to estimate conditional probability could be replaced by subtraction. At the time the device was constructed, digital computers were rare and expensive, but apart from that it was believed that an analog solution was likely to have closer correspondence to events in neural circuitry. This argument was strengthened by the ease with which a "forgetting" feature could be introduced, simply by connecting resistors across the storage capacitors to allow their "counts" to drift towards zero. The inferences made by the computer could then be seen to be more strongly influenced by recent inputs than by earlier experience, conforming to human and animal behaviour.

The conditional probability computer could be made to simulate some forms of animal learning. It had no intrinsic appreciation of reward or punishment, so an indication of a favourable response had to be given. The device was sometimes demonstrated as modelling the training of a dog, where the five input channels were labelled so that two of them (b and s, say) represented the commands "beg" and "sit" and another two (B and S) represented corresponding actions by the dog, and the fifth (r) represented a reward in the form of a bone.

The dog was "trained" by presenting the inputs bBr, bS, sSr, sB repeatedly in random order. It would then make the appropriate response (as an inference) to a command (B to command b, and S to command s) provided it was "made hungry" by being "shown the bone". That is to say, there would be an inference B in response to the input br, or S in response to sr, though no inference to inputs of b or s alone.

The need to "show the bone" does not invalidate the scheme as a model of biological learning because a tendency to seek particular inputs can be seen as a consequence of the needs of survival. Nevertheless the demonstration with discrete input channels, in common with other models or accounts of classic and operant conditioning, does not provide the required starting point for the examination of intelligence. This is because the inputs depend on complex processes underlying pattern classification. The dog modelled in the above example has to be able to distinguish spoken words as well as the image of the bone, and in reports of classic conditioning it is assumed the subject can differentiate between the sound of a bell and for example that of a whistle. A familiar human example is the feeling of appetite that is aroused by quite specific sound patterns that customarily precede eating, such as the rattle of teacups.

However, although the phenomenon of conditioning is usually studied and described in ways that require the participation of complex processes of sensory analysis, appreciation of correlation is the basis, and could become effective at a more primitive level. This would be such that pattern classification is not required and the organism responds to correlation between continuous variables. This seems to be an appropriate starting point for the study of intelligence, with pattern classification and concept-based processing emerging as later evolutionary developments.

This viewpoint was reached after an attempt to apply the principles of the conditional probability computer in a practical control situation.

Application to Process Control

The problem of reconciling continuous and conceptual information processing arises in attempts to make useful "learning machines" or self-improving devices to perform practical tasks. There is an implicit imprecision in any term (such as "self-improving" or "self-organising") that implies that an entity modifies itself. Whether the entity is seen as self-improving, or usefully self-organising, must depend on how it is described, as there has to be a mode of description according to which the purported improvement was included from the start. There is a problem in identifying the "self" that organises. The modification, of course, is not expected to occur in isolation, but in interaction with an environment.

A useful viewpoint is due to Glushkov (1966) who describes a self-organising system as being decomposable, at least notionally, into a learning automaton and an operative automaton. The latter determines the behaviour of the system as tested and described (for example, the response of a person to the posing of a question). The learning automaton is the part of the system that modifies the operative automaton as experience is gained. The division between the two must be somewhat arbitrary and may depend on the length of time over which the behaviour attributed to the operative automaton is observed. An interesting point is that Glushkov considers the possibility of more than two levels of automata, so that there could be a learning-to-learn automaton that would operate to improve the operation of the learning automaton, and a learning-to-learn-to-learn one, and so on. (The term "learning-to-learn" appears in the psychology literature, and biologists are very ready to account for aspects of animal behaviour, or acceptance of pollination by plants, as favouring appropriate levels of genetic diversity, and hence as evidence for "adaptation-to-adapt".)

Despite these complexities, there is fairly general agreement about what is meant by "learning" as a characteristic of people and animals, and as something that it is useful to imitate in artifacts. At a relatively simple level this can amount to "tuning" or automatic adjustment of numerical parameters, for instance of the gains of the three components in a PID servo mechanism. The adjustment would be made with respect to a suitable criterion of fidelity with which the servo output conforms to its input. Such adjustment can be considered a simple form of learning, although Pask (1962; see Andrew 2001a) insisted that the term should be reserved for processes that altered the representation or function more drastically.

Pask's insistence was that learning should necessarily involve what is discussed below as "classification". The relatively simple process of "tuning" or parameter-adjustment was termed by him "adaptation" However, it will be argued here (a) that "tuning" can be a much more complex process than at first may appear, and (b) that there is a route by which classification can arise as a development of "tuning", and that this offers a highly plausible theory of the origins of biological intelligence. "Classification" can be roughly equated to concept-based processing, though in a primitive form where a concept need not be named in any language external to the learning organism.

Learning machines operating to adjust continuous variables have been widely assumed to be uninteresting as models of learning, and there have been several efforts to adapt schemes having discrete inputs and outputs. One such scheme is the conditional probability computer due to Uttley, already described as a means of modelling conditioning. The desire to introduce these discrete methods was prompted by the fact that human control is at least partly concept-based. It does not, however, follow that discrete operation should play a part from the start. According to the viewpoint that will be developed here, attention should first be given to continuous control, from which classification and concept-based processing can emerge.

An adaptive controller applied to a continuous industrial process is shown in Fig. 2.1. The controller receives signals a, b, and c from transducers in the process plant and generates control signals d and e. Associated with the controller is a means of evaluating the degree of goal achievement, normally an estimate of profit for an industrial process though other considerations might also enter. This is labelled h for "hedony". The controller is required to find a way of computing d and e from the inputs so as to maximise h.

Boxes

One way of making the continuous signals acceptable to a controller accepting discrete inputs is to divide the range of possible values from each transducer into a number of intervals and to let all values within the interval be equivalent. This means that the multidimensional space (Three-dimensional if there are three transducers) in

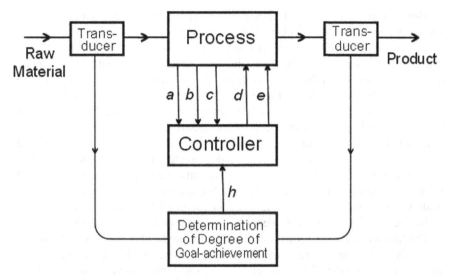

Fig. 2.1 Process with adaptive controller

which a point represents the state of the process (insofar as it is monitored by these transducers) is divided into rectangular "boxes" within each of which all points are equivalent as far as control action is concerned. There will usually be a large loss of information in going from the continuous to the discrete representation, but this may be acceptable.

Michie and Chambers (1967) developed a system they called BOXES to explore adaptive control. One of the applications they described was learning to play noughts-and-crosses (tic-tac-toe) by machine, though because this operates in a discrete environment, the special feature of division of a continuous representation space into "boxes" was not involved. They also considered a control task of a continuous nature, namely the balancing of an inverted pendulum, or pivoted pole, attached to a driven trolley. (This has also been used as a sample task for experiments in machine learning by Donaldson [1960, 1964], Widrow and Smith [1964], and Barto *et al.* [1983], but using quite different approaches.)

In the experiments of Michie and Chambers the pole, or inverted pendulum, was simulated in a computer program, but it is convenient to describe it as a physical mechanism. Like a clock pendulum, it was constrained to move in a plane. The pivot at its lower end attached it to a motor-driven trolley running in the same plane on a length of track. The task to be learned was that of controlling the motor of the trolley so as to keep the pendulum from falling over while also not allowing the trolley to run off either end of the track.

Signals made available to the controller, as though coming from attached transducers, indicated the position of the trolley along the track, and the angle of the pendulum, and the time derivatives of each of these. The range of the angle was divided into six intervals, and the ranges of the other three variables into three intervals each, so that the total number of "boxes" was $6 \times 3 \times 3 \times 3$, or 162.

The motor driving the trolley operated in what is termed "bang-bang" fashion, being always fully energised in one or the other direction. Time was suitably quantised, and at each step a decision was made as to whether the motor would drive to the left or to the right in the next interval.

A pair of storage locations was associated with each of the 162 boxes. In one of them was stored a value for the mean time to failure after a decision to drive to the left when the point indicating the current state fell inside the box. The other stored a value for the mean time to failure after a decision to drive to the right. The values were "smoothed", or weighted in favour of recent experience. In the early stage of learning the decisions were made randomly, but as experience was gained they were made to favour the choice that was associated with a longer mean time to failure. Failure occurred when either the pendulum fell over or the trolley ran off the track.

The task as described, as well as that of learning to play noughts-and-crosses, differs from the situation represented in Fig. 2.1 in having to wait for the end of an episode of operation before receiving a signal corresponding to h, or hedony. Once the system has learned to keep the pole from falling, and the trolley from leaving the track, there is no further change in control policy. It would of course be a simple matter to include in the hedony feedback an extra component allowing reinforcement of policies that minimise the deviations of the pole from its upright position.

As a general model of learning in a continuous environment, the BOXES scheme has obvious drawbacks. It depends on partitioning of the operational space in an arbitrary way, whose effectiveness only becomes apparent in retrospect. A possibility that could be considered would be to let the partitioning be modified adaptively as learning proceeds, such that adjacent boxes might be merged if the feedback signals to each were found to be similar, and boxes whose feedback signals were particularly variable might be split.

In any version, however, if the partitioning is too coarse, there must be a large loss of information in going from continuous values to the discrete form, and this almost certainly means that the quality of control is less than what other techniques would allow. On the other hand, if the divisions are too fine, the number of "boxes" is increased and can readily become unmanageable. In the limit the boxes could be made to correspond to the limits of discrimination of the transducers, so that there would be a separate box for every discriminable state of the system. The penalty for this is not only the increased cost of the necessary storage, and of complexity of the means of updating it, but more significantly in the time needed for learning. Learning a control policy would require that the point representing the state of the system entered each of the boxes a substantial number of times, so as to allow the collection of statistical data relevant to it.

The number of discriminable states of the system, equal to the product of the numbers for all the monitored variables, is such that the time needed for learning in this way is prohibitive in any practical situation. The conclusion has to be that continuous variables have to be treated as such, so that interpolation and extrapolation are possible. These clearly play a part in human and animal learning, as can be seen for example by considering a person learning to ride a bicycle. It is unimaginable that the person would know what to do when the bicycle had tilted, say, three degrees from the vertical, and also five degrees, and yet have no idea what to do for a tilt of four degrees.

An obvious disadvantage of the BOXES method, if used to generate a continuous output, is the inevitable discontinuity in crossing a partition between boxes. A smoother output could be obtained by associating the collected statistics with an array of key points distributed throughout the state space, for instance with a key point at the centre of each "box". Then the control output for any state of the system could be generated by interpolation from the key points nearest to its representation in the space. Such a method takes account of continuity and is classifiable under the heading of "learning filters" discussed in the next section.

Fuzzy set theory, as introduced by Zadeh (1965), recognises the need to combine continuous and concept-based forms of processing. It has allowed new and valuable approaches to adaptive control, as treated by Pedrycz (1989) and in a vast literature in journals and conference.

The fuzzy approach combines the discrete and continuous in a way that is opposite to the evolutionary viewpoint advocated here. The fuzzy approach accepts concept-based processing as a basis and extends it by allowing concepts to be "fuzzy". An example is the concept of a "tall man", which is imprecise because there is no exact threshold of height beyond which a man is recognised as being tall. Zadeh showed that such "linguistic variables" can be treated effectively in terms of membership of fuzzy sets. In this particular case, the degree of set membership is a function of the actual height, rising from zero at some height (say 1.7 metres), which would definitely not be considered tall, to unity for heights of, say, 1.9 metres and above.

Because fuzzy techniques assume concept-based processing, they do not seem to provide the required starting point for the examination of intelligence. Their practical value is indisputable but is realised in combination with human judgement. Their application to adaptive control is particularly valuable in nonlinear situations, but, as with the use of BOXES, they generally depend on manual setting of parameters. Examples are the upper and lower limits mentioned above for "tallness". Fuzzy relationships are usually represented by concatenations of straight-line segments, the x-coordinates of whose junction points are somewhat arbitrary.

Learning Filters

Adaptive controllers treating continuous variables as such have received a great deal of attention. An early example was by Draper and Li (1951) who applied such a controller to adjust the ignition timing and the proportions of the fuel/air mixture of an internal combustion engine so as to maximise the power developed. One general approach to adaptive control is referred to as Kalman filtering (Sorenson 1985; Visser and Molenaar 1988). Methods for optimisation of computer models are reviewed by Schwefel (1981) and these are generally applicable also to real-time online optimisation. The topic is discussed further in Chapter 4.

An important point that will be developed in Chapter 4 is that the changes possible within an adaptive controller, as represented in Fig. 2.1, can include selection of terms as well as adjustment of numerical parameters. One possibility is to let the control signals d and e be computed as polynomial functions of the sensed variables a, b, and c, such as:

$$d = K + La + Mb + Nc + \text{higher-order terms}$$

where the parameters K, L, M, N, and others appearing in the higher-order terms are adjusted during operation so as to maximise h. The adjustments can be determined by fairly simple statistical estimates discussed in Chapter 4, though the situation is complicated by influences on h that are not connected with the control action, referred to as noise, and also the uncertain time delay between a control action and its effect on h.

The means of adjusting the polynomial can readily be extended to allow changes in the set of terms included in the polynomial. These can be seen as changes in the form of the polynomial, and of the connection pattern of a set of neuron-like elements continuously evaluating it as the control signal d. A controller working in this way can merit description as self-organising, which is understood to imply the possibility of fundamental changes in structure. The meaning to be attached to "fundamental" is not precise, as for example in a network the making of a new connection is functionally equivalent to turning a gain setting from zero to a nonzero value. Description as self-organising is probably warranted when the number of connections effective at any one time is small compared to the number of possibilities.

Certain measures of correlation can indicate the desirability of introducing new terms to the polynomial. (An alternative would be to introduce new terms randomly "on approval" and then to eliminate a term if its coefficient remains close to zero for a sufficient time.) The introduction of terms of degree higher than one allows for nonlinearity. The method can also allow the introduction of new variables, additional to the a, b, c, and so forth, initially present. This shows correspondence to the establishment of the conditioned reflex, where a new signal such as the sound of a bell or whistle comes to be associated with the unconditioned reflex linking the appearance of food with salivation.

The manipulation of a polynomial in this way has much in common with recommendations for the forming of regression models, for instance by Draper and Smith (1966). There is however the important difference that the task of regression analysis has the characteristics of "learning with a teacher". That is to say, a "correct answer" is made available to the learning system, so that the magnitude and sign of error can be determined. Any system required to form a model or to make a prediction has this character. A system required to optimise its control with only an indication of hedony, or degree of goal achievement, has to operate differently, as will be discussed in Chapter 4. The difference is sometimes indicated by reference to optimisation with odd and even objective functions.

The use of a polynomial is not the only way of representing a continuous relationship between variables. The possible use of a grid of "key points" in the multi-dimensional space defined by the monitored variables has been mentioned. A kind of self-organisation is possible here, too, allowing modification of the key point locations. This could operate by eliminating a point from the grid if its stored values were close to what could be derived by extrapolation from neighbouring points, and using other criteria to indicate where the density of points should be increased. There are possibilities to be explored here, but use of a polynomial, as is also customary in regression analysis, seems more convenient. A particular disadvantage of the key-point method is that the introduction of new variables requires a change in dimensionality of the representation space. Artificial neural nets offer yet other methods for representing continuous relationships in ways that allow adaptation.

Possibilities for self-organising controllers operating on continuous variables were discussed by Andrew (1959b) as the outcome of an attempt to apply Uttley's discrete-event conditional probability computer to process control. At the time the discrete-event and continuous approaches were seen as irreconcilable alternatives, but the view of conditioning as a response to observed correlation allows a unified view encompassing both.

A similar scheme was implemented in hardware by Gabor *et al.* (1961) and termed a "learning filter", a term that has come to be applied generally to schemes of the kind. Strangely, although they referred to the device as "learning", Gabor and his colleagues denied that it had any relation to human intelligence. This was in reply to a question from Gordon Pask following the spoken presentation of the paper (in discussion appended to the printed text). Instead, these authors stressed the classificatory aspect of human mental activity and claim that an attempt to model human intelligence should embody this from the start. They are of course right in claiming that classification is important, but not to the exclusion of other functions.

Although not usually recognised as such, the famous checker-playing program of Samuel (1963) incorporates what is really a continuous-variable "learning filter". This is probably the most famous learning algorithm to emerge in Artificial Intelligence and although Samuel worked on checkers, or draughts, his methods have been carried over into programs to play other games including chess. That a learning filter finds application in these totally discrete task environments lends support to the suggestion that something of the sort participates in biological intelligence.

The making of a move by one player in chess typically requires a choice from about 30 legal possibilities. The number of possible courses the game might take in only a few moves is enormous, as each of the 30 moves could be met by any of about 30 responses from the opponent, and so on. A skilled chess player can look at the configuration of pieces on the board and can say something about the relative strengths of the two players' positions. There is therefore interest in forming a "static evaluation function", sometimes called the "scoring polynomial", to indicate the relative strength of a player's position.

Once a static evaluation function has been devised, the making of a move should be simple, as it is only necessary to choose which of the 30 alternatives leads to the most favourable outcome according to this criterion. However, the game of chess, or checkers, is not so easily mastered, because no one has devised a static evaluation function that produces powerful play. The adjective "static" indicates that the function is evaluated without exploration of possible game continuations (i.e., without lookahead). The ideal method of play in which every possible course of the game is followed to completion is easily shown to be infeasible.

Static evaluation functions have been employed by chess players since well before the days of computer chess. Where a game has had to be abandoned without hope of resumption (which, in the romantic tradition of chess, would presumably be because an enemy was at the gates of the city, or one or both players was about to go to the gallows), such functions have been used to decide a winner.

It has been found that the quality of play is increased greatly if the static evaluation is applied after the tree of possible continuations has been explored to some depth. Once the evaluation has been made at each of the final nodes, or "leaves" of the tree, values can be fed back up through the tree, with the assumption that each player, at each stage, makes the choice most favourable to him. That is to say, if the exploration is made when player A is next to move, and the static evaluation function is computed so as to represent advantage to A, at each node of the tree that represents a nonfinal move by A the value is taken from the daughter node with highest value, whereas at nodes corresponding to a move by the opponent B the value is taken from the daughter node having lowest value. The feeding-back of values through the tree is referred to as minimaxing.

With a branching factor of about 30 at each node, it is obvious that the lookahead tree proliferates rapidly. Much ingenuity has been applied to reducing the complexity without compromising the effectiveness of the process, for instance by restricting attention to moves that are, by some criterion, plausible, or by recognising board configurations as turbulent or quiescent, with preference for the latter as final "leaves" at which static evaluation is applied. More valuable than either of these is the use of "alpha-beta technique" as a strategy for exploring the lookahead tree and performing the necessary minimax operation on it. The order of evaluation of nodes can be such that large parts of the tree can safely be ignored and the computation time correspondingly reduced. These methods can be referred to "as pruning the tree". Plausible move selection and the like represent forward pruning, and alpha-beta technique is referred to as backward pruning.

The main point for the current purpose, however, is that the essentially discrete game situation is usefully represented by a numerical "scoring polynomial", which, if of first degree, is simply a weighted sum. The variables in this are numerical estimates of features of the play position well known to experts, of which the most obvious is termed "piece advantage" or "material balance". This is simply a comparison of the pieces still held by each player, the pieces being suitably valued. There is no point in attaching values to the kings for this purpose, as the fact that the game has not ended means that both kings are on board. The highest value is given to the queens, and then considerably lower ones to knights, bishops, and rooks, and the lowest of all to pawns.

Other variables in the scoring polynomial (at least, as set up by Samuel for checkers) include "mobility", of which a rough indication can be formed by comparing the number of legal moves open to each player, and "centre control". Samuel's primary interest was in machine learning, and he arranged for the coefficients in the polynomial (corresponding to K, L, M in the earlier case) to be adjusted, or "tuned" as experience was gained, and he also included means of modifying the selection of terms.

A difficulty is that there is no suitable feedback of the degree of goal achievement, or hedony, to guide the adjustments. In the BOXES approach to learning to play noughts-and-crosses (tic-tac-toe), all adjustments were deferred till the end of the game, and then counts associated with each of the possible play situations were updated according to whether the game was won, lost, or drawn. This is possible because for noughts-and-crosses the number of possible play situations is relatively small, but for checkers or chess the number is enormous.

For his checker-playing program, Samuel overcame the difficulty by exploiting the fact that the quality of play of a program improves with increase in the depth of lookahead. The greater the depth, the less the performance depends on the exact form of the static evaluation function (s.e.f.). This shows that the evaluation of a situation obtained by backing-up from a lookahead tree is more effective than that obtained by applying the s.e.f. directly. The learning algorithm devised by Samuel operates to reduce the difference between the backed-up evaluation and the evaluation formed directly. It therefore operates with an odd objective function because the difference between the two values is similar to an indication of error. One snag is that perfect agreement between the two values could be obtained by setting all of the coefficients to zero. To prevent the program finding this unwanted minimum, it was arranged that the piece-advantage term was always present with a fixed coefficient.

Samuel's program was successful in improving its own playing ability in checkers, sometimes by playing repeatedly against another version of itself in the same computer. It is particularly interesting to consider the internal workings of game-playing programs in connection with an early debate between Ashby and Bellman after an address by Ashby (1964) in which he advocated a vigorous program of research on computer chess as a way of investigating intelligence in general. More will be said about this in the next chapter.

Classification versus Tuning

The schemes that have been discussed under the headings of process control and learning filters all refer to learning devices dedicated to one particular task. On the other hand, a person or animal faces different tasks and environments at different times, and something that looks like a single task from one point of view may drastically alter in character. An example of the latter is the task of walking upright, which has one character when the feet make nonslipping contact with the ground, but another when slithering on black ice, and similar considerations apply to driving a vehicle. The essentially continuous "learning filter" or "tuning" process needs to be supplemented by classification of task environments, with distinct sets of "tuning" parameters, defining control policies, for the respective environments.

For people and higher animals, the task situation may change because of internal decisions to pursue different goals, but at a primitive level the changes are likely to be imposed from outside. A means of classifying environments, so as to bring an appropriate "learning filter" into play for each, has somehow been evolved. The classification process is presumably invoked in response to a failure of a biological "learning filter" to converge, especially when the failure follows a particular pattern. A pattern that would strongly suggest a multiplicity of environments, and hence a need for classification, would be one where for a time the learning filter appeared to be converging on a policy and then abruptly began to diverge from it.

Details of how such evolution might have happened can of course be only speculative, but some such mechanism has to be visualised as underlying the transition from continuous-variable processing to classification and hence concept-based processing.

Nontrivial Machines

MacKay (1959) introduced a version of information theory in which he distinguished *logons*, or units of structural information, from *metrons*, or units of metrical information. As will be argued in the next chapter, the sensory inputs to a biological learning system are metrical, and the system derives structure from them. A basis for a possible mechanism was indicated in the foregoing section.

There is also a connection with the early work of Pask (1962; see Andrew 2001a) in which he maintained that any process to be described as "learning" must introduce a new descriptive language. Anything not having this feature he termed "adaptation". The reference to "language" is in a special sense discussed by Arbib (1970) and does not imply the syntactic structure of a spoken language. The intention is to focus on fundamental changes of representation such as that from continuous representation to classification.

At the much more advanced level of human behaviour, it is possible to see evidence of innovation prompted by confrontation, in a way that is reminiscent of the postulated emergence of classification from opposing influences on a learning filter. Von Foerster (von Foerster and Poerksen 2002) has described humans as nontrivial machines. One illustration of this (not his) is that a person, shown the sentence "This sentence is false", does not go into an endless loop of decisions ("true, therefore false, therefore true, therefore false ...") as a trivial reasoning machine might. Instead of being trapped in such a loop a person classifies the situation, either inventing a term to describe it or recalling the existing term "paradox".

Conclusion

Out of a variety of possible starting points for the study and analysis of intelligence, it has been argued that the most valuable is the examination of schemes operating adaptively on continuous variables. The importance of continuous operation will be defended from additional viewpoints in the next chapter. Concept-based processing is also extremely important, to the extent that facility in employing it should probably be seen as the main feature that gives humans a claim to mental superiority over other creatures. However, it should be seen as evolving from continuous operation, and suggestions are made as to how this could have come about.

The importance of continuous processing is defended on evolutionary grounds. This does not mean that it has to be apparent in all manifestations of intelligence. People are able to operate effectively in many areas without conscious reference to continuity, using faculties that have been shaped by billions of years of evolution. On the other hand, evolutionary developments usually leave traces of earlier forms, and it would be surprising if there were not remnants of primitive continuous processing operating subconsciously.

My own feeling is that the key to further progress in understanding intelligence, and to breaking the current impasse in AI, is to be found by examining the interface between continuity and conceptual processing. To some extent this is just a "hunch", but arguments will be developed later, especially the discussion of the fractal nature of intelligence in Chapter 7, that lend support.

The term "continuity" has been used without definition, relying on intuitive interpretation. The distinction between discrete and continuous representations is treated by Klir and Elias (2003). As they point out, any observation of a value, whether by a physical instrument or a living sense-organ, is associated with some unavoidable finite error. It follows that the resolution level is limited, and as they say: "This implies that data are always discrete regardless of philosophical beliefs or the state of technology".

In dealing with observed data, continuity "in the small" can have no meaning except as part of a constructed theory. By this is meant continuity as demanded by differential calculus and defined by saying that a function $f(x)$ is continuous at α if its limit, as $x \to \alpha$, is $f(\alpha)$. For the discussion here, continuity has to be interpreted "in the large", implying linear ordering and recognition of distance. These are the necessary conditions to allow interpolation, extrapolation, and smoothing, of which at least the first is important in the many forms of learning that constitute skill acquisition.

Klir and Elias also observe that the advent of digital computers has encouraged the use of discrete rather than continuous methods of analysis. This of course is a comment on present-day practice rather than on the origins of intelligence, and also it remains true that most if not all digital computing makes some use of a number system, and hence of linear ordering and recognition of distance.

Summary of Chapter 2

A number of possible starting points are considered for the description and study of "understanding" or "intelligence". These include the rather usual assumption that discrete logic is fundamental, which finds expression in discussions by

McCulloch and Pitts and in the Laws of Form of Spencer Brown. Another candidate is a capacity for associative recall, which can be implemented in artificial systems using either Hopfield neural nets or analogues of holography. Each of these embodies an assumption that useful concepts have already been formed so must be underpinned by something more basic.

Neuroscientists have discussed the "binding problem" in perception, usually with an assumption that frequently occurring patterns of excitation come to be recognised. The emphasis on frequency is not consistent with a view of evolution as guided by survival value, both in determining binding and in the evolution of sense organs. Modelling of the environment has also been suggested as a fundamental property of an intelligent system, but it is argued there is no clear distinction between a model and a policy for interaction.

A useful starting point is provided by the study of conditioning, particularly operant conditioning in which the response to be conditioned is rewarded, typically by a food pellet. As usually described, this does not offer a starting point because it depends on a pre existing capacity for discrimination of complex stimuli. However, the basic principle of response to correlation between something external to the organism and something beneficial can refer to continuous variables and can be the basis of a continuous adaptive controller of the "learning filter" type. This can embody means of term selection that can be seen as self-organisation, but in order to operate in different environments it must be supplemented with a means of classification of environments. Such classification constitutes the beginning of concept-based processing and consideration is given to how it could arise. There is a connection with early theories of Pask and with von Foerster's view of people as nontrivial machines.

Chapter 3
Continuous versus Discrete

The Continuous Environment

The environment, as people perceive it, is essentially continuous, and we are conscious of moving and interacting in the three dimensions of space, along with that of time. The same is true of many important derived variables such as mass, liquid volume, light intensity, and so on. In some cases the apparent continuity is known to break down at the atomic level, as for example a volume of liquid consists of discrete molecules, and light is made up of photons, and it is possible that other variables such as length and time may be quantised at some level. Nevertheless, as far as direct perception is concerned, everything seems to conform to old-style elementary physics in which variables are treated as continuous. It is reasonable to suppose that evolution, including that of intellect, occurred in an environment that is either essentially continuous or has been made to appear so by evolutionary choice of the means of experiencing it.

An early and vigorous advocate of the importance of continuity as an aspect of intellect was Donald MacKay (1959), whose views are discussed at some length in interviews reported by McCorduck (1979). He refers to various ways in which continuous variables can be argued to play a part in intelligent activity, including weighting of conflicting bodies of evidence, and estimates of distance from a solution in problem solving. These are treated further in Chapters 6 and 7, along with the related idea of "heuristic connection" between problems, postulated by Minsky (1959, 1963).

MacKay accepts that the estimate of "distance from a solution" may be simple and obvious in certain cases but insists that this is exceptional. He refers to the solving of a jigsaw puzzle, where the amount of unfilled space could be a simple criterion of such distance. (Even here, the simple criterion is not altogether appropriate because the difficulty of filling a contiguous area depends on the amount of structure in that part of the picture. The completion of areas of plain blue sky is notoriously difficult. It can also be argued that a linear function of the unfilled area is not the most appropriate measure of difficulty.)

Broadly speaking, work under the heading of AI tends to de-emphasize continuity. Some of its essence is captured by the use of the relational operators "less than",

A.M. Andrew, *A Missing Link in Cybernetics*, IFSR International Series on Systems Science and Engineering 26, DOI 10.1007/978-0-387-75164-1_3, "equals", and "greater than", and because these give "true" or "false" outcomes they fit well into the "production rules" and other formulations of mainstream AI. However, they do not capture all the implications of continuity and are often employed with reference to arbitrarily chosen fixed values. If these values are adjusted in the light of experience, either manually or by a learning algorithm, the primacy of continuous variables has been tacitly acknowledged.

When referred to time, the relational operators become criteria of precedence and simultaneity, and as such certainly do not fully represent the familiar conception of time, epitomised as "time like an ever-rolling stream". (Einstein's theories of relativity show that precedence and simultaneity become rather complex issues for observers using different frames of reference, but in the presumed conditions of biological evolution relative velocities are minute fractions of the velocity of light and again a simple "old-style physics" view is permissible.) Rather curiously, although "time as an ever-rolling stream" is difficult to define, it can be measured with unimaginable accuracy by atomic clocks, of which the best actually has a fractional error between 10^{-14} and 10^{-13}. Impressive accuracy, though falling short of this by many orders of magnitude, is achieved even by cheap digital watches containing quartz crystals.

The importance of continuity is also acknowledged by Ashby (1956, 1960), in his graphical representation of "state-determined systems". The current state of a system is represented by a point in multidimensional space (for illustrative convenience usually shown two-dimensional). Such representation acknowledges continuity, with system behaviour represented by a continuous trajectory. The tabular representation of a finite automaton as used in many theoretical treatments of computation (Minsky 1967) also defines a system that is state-determined but that takes no account of continuity. (The finite automation formulation normally assumes the occurrence of input signals to trigger and partially determine state transitions, but these can be replaced or supplemented by internally generated clock pulses.)

There is a trend in modern mathematics and theoretical treatments to start from a basis of set theory and to introduce continuity as a special feature, where appropriate. This has the methodological advantage that theory can be developed that is applicable with or without continuity. Set theory is used in this way by Ashby and by Klir (Klir and Elias 2003; Klir 2006). The definition of a "function" in a textbook adopting the modern idiom (Jeffrey 1969) is as follows:

> The modern definition of a function in the context of real numbers is that it is a relationship, usually a formula, by which a correspondence is established between two sets A (the domain) and B (the range) of real numbers in such a manner that to every number in set A there corresponds only one number in set B.

However, the view that has proved most useful in modern mathematics need not correspond to the intuitive feeling that most people have about physical variables impinging on them, nor to the way these have influenced evolution. Set theory was, in fact, slow in gaining acceptance even among mathematicians, and it is easy to believe that the view of continuity that led Newton and others to devise the differential and integral calculus had a different flavour.

Catastrophe Theory and Dissipative Structures

Even when some form of continuity is accepted, there tends to be an assumption that the natural numbers are in some sense fundamental or God-given and that ingenuity has to be exercised to fill the gaps between them, first with rational numbers and then others. From an evolutionary viewpoint, it is more natural to think of continuous variables as fundamental and to consider how the natural numbers came to be appreciated. Catastrophe Theory (Zeeman 1977; Thom 1983) indicates one bridge between continuous phenomena and discrete events, and some of the discussion associated with it gives support for the suggestion that continuous variables should be seen as fundamental.

The work of Prigogine and his followers (Prigogine and Stengers 1984) on thermodynamics also indicates such a bridge, as it is shown that in "dissipative structures" through which there is a flow of solar or other energy, order can arise spontaneously.

It is very likely that the kinds of event postulated within both catastrophe theory and theory of dissipative structures have played a part, within living organisms, in the evolution of concept-based processing. They produce order or structure, but without regard to any benefit, and consequently can only be part of the story.

Corning (2005) has pointed out that in much discussion of self-organisation there is a failure to distinguish two radically different kinds. One of them is nonpurposive in nature and he suggests it should be termed "self-ordering", whereas the other is ends-directed. Evolution almost certainly depends on self-ordering processes, but only in conjunction with selective mechanisms that give direction.

Nervous System

There is no doubt that parts, at least, of the nervous system are concerned with processing continuous data. Electrophysiological recording of nerve action potentials usually shows signals that are characteristic of frequency modulation of impulses. Inputs from sense organs indicate the intensity of various kinds of stimulus, and particularly their rates of change, in terms of impulse frequency, and outgoing signals to muscles are similarly coded. Activity in intermediate neurons, not directly associated with either input or output, has a similar character.

Biological motor control and the regulation of many internal variables of the body (the *milieu intérieur*) are readily accepted as depending on manipulation of continuous, or analog, variables. The extreme precision that can be achieved in motor control is evident from the performance of various kinds of sportspeople, and a high degree is also demanded by everyday activities of moving around and handling objects and by stabilisation of the retinal image. It is clear that the nervous system is capable of processing continuous data, and it would be surprising if this capability was distinct from those underlying intelligence. The regulation of internal variables depends at least partly on nervous control, as acknowledged by reference to part of the nervous system as autonomic.

One, among very many, of the internal variables that are closely controlled is the core body temperature in humans. This remains stable despite wide variations in ambient conditions. It was shown by Cornew, Houk, and Stark (1967) that the regulation operates at two levels. The temperature of peripheral parts of the body is controlled relatively crudely by such mechanisms as sweating and shivering, and then fine control is achieved by regulation of the blood vessels that determine the exchange of heat between periphery and core.

Another regulatory loop that has been much studied experimentally is that controlling the size of the pupil of the eye and hence the amount of light reaching the retina (Stark 1959; Bayliss 1967). This has the great advantage, from the point of view of experimentation, that the effective gain of the loop can be altered, in either direction, by means other than surgery. (The gain may be reduced by arranging that light enters the pupil centrally, so that small changes in pupil size have little effect. Conversely, it can be increased by arranging that most of the light enters near the edge. The main means by which eyes adjust to widely varying levels of light is by change of retinal sensitivity, supplemented by the pupillary reflex in providing rapid response.)

Biological regulation depends on other mechanisms besides neural activity, as shown by its occurrence in organisms with no nervous systems at all, including single cells, but neural circuits play a large part in its achievement in higher animals, providing further evidence that neural circuits are well suited to processing continuous signals.

The model neurons due to McCulloch and Pitts (1943) are essentially "logical" elements, in the sense of discrete or "computer" logic. In the original formulation, each excitatory synapse has weight unity and excitation thresholds are small integers, and inhibitory synapses are given the power of absolute veto. It is interesting that almost all of the types of artificial neural network that have proved useful in recent decades, starting with the perceptron due to Rosenblatt (1961), have departed from the formulation in having continuously variable synaptic weights. It does not follow that real nervous systems should be expected to have this character, but it gives grounds for supposing that they may. The continuously variable weights allow incremental learning.

The observation that the activity of neurons is characteristic of transmission of continuous information has to be reconciled with the fact that people are capable of thought using discrete concepts. At a more phylogenetically primitive level, the reticular formation of the brain stem (Kilmer, McCulloch, and Blum 1968; Scheibel and Scheibel 1968) has been interpreted as a means of switching the animal among a repertoire of discrete states. The behaviour of quite primitive animals, such as that of the earthworm in pulling leaf-like objects into its burrow (Darwin, 1897), is difficult to describe except in terms of recognition of discrete concepts. All-or-nothing responses are implicit in such discrete or concept-based behaviour and could be produced by positive feedback in a mass of continuously acting neurons. This would correspond to the familiar achievement in electronics of bistable, or "flip-flop" or "multivibrator" behaviour by applying positive feedback to an amplifier (transistor or vacuum tube) that could otherwise accept continuous input. A mass of neurons reacting in this way may correspond to a cell assembly as postulated by Hebb (1949).

It is in fact possible to see a curious alternation of discrete and continuous features in neural evolution. Maintenance of a living cell has to depend on continuous chemical processes, but in the special case of neurons an all-or-nothing discrete response has emerged. This in turn has a generally continuous character as far as information transmission is concerned but appears to be used in "flip-flop" fashion to again give discrete behaviour. The discrete concepts then formed become in turn the basis of continuous measures of the type discussed below under headings of *heuristic connection* and *basic learning heuristic*. A somewhat similar interplay of continuous and discrete features occurs in evolution considered generally, the latter represented by the genetic code, which has complex interactions with the continuous structure and function of organisms.

Evolution and Learning

Another reason for interest in continuity comes from the observation that any process of evolution or of spontaneous (i.e., without a teacher) learning must depend on it. Lenat (1984) acknowledges the dependence of evolution on continuity in an AI context, with the assertion that "Appropriateness (action, situation) must be a continuous function." He quotes a similar assertion by the mathematician Poincaré referring to analogical reasoning.

There are in fact two ways (as well as a third, noted below) in which continuity enters into evolution or spontaneous learning. One is the part played by continuous criteria of similarity of situation and of solution as required by the *basic learning heuristic* postulated by Minsky and Selfridge (1961):

> In a novel situation try methods like those that have worked best in similar situations.

At first sight this seems to be too obvious to be worth stating, but two continuous criteria of similarity are invoked, namely similarity of situation and similarity of method. In many task environments, suitable choices for these are far from obvious. This is so, for example, in the acquisition of skill in a game such as chess, where the criterion of similarity of situation has the intangible character of the *heuristic connection* between problems postulated elsewhere by Minsky (1959b, 1963), which will be discussed further in Chapters 6 and 7.

The heuristic can also be seen as operating in simple tasks of learning or optimisation. In learning to ride a bicycle, for example, advantage will certainly be taken of the fact that the action to be taken when the bicycle and rider have tilted 5 degrees to the right is probably similar to that for 4 degrees, namely to steer to the right to restore the upright position. Continuity cannot be guaranteed everywhere, and there is certainly an angle of tilt beyond which recovery by steering is impossible and the rider must either put a foot to the ground or prepare to contact it more ignominiously. Within a certain range, continuity applies and it is unthinkable that a rider would have learned what to do for tilts of 3 and 5 degrees but have no idea how to react to 4 degrees. Exploitation of measures of similarity allows evolving or learning systems to build on earlier progress.

The exploitation of similarity (of situation and of method) is one of the ways in which evolution or learning can utilise continuity. It does not necessarily imply a continuous mapping between these two, though often this will exist and can also be exploited, and is a second way of using continuity. For example, the above-mentioned person learning to ride a bicycle, and knowing what to do for tilts of 3 and 5 degrees, may guess, not only that the action for 4 degrees should be similar, but that it is probably about half-way between the actions for 3 and 5 degrees. This use of interpolation implies an assumption of a probable continuous mapping, and the assumption may also be used to justify extrapolation, such that a learner knowing what to do for 3 and for 4 degrees can make a guess (which may be wrong) at what to do for 5 degrees.

The need to exploit continuity in a learning process can be illustrated by returning to the learning or self-optimising controller for an industrial process, discussed in Chapter 2. It may be assumed that the control action to be taken depends on signals from half-a-dozen transducers in the process plant. If the outputs of each of these are divided into 20 intervals, the number of states that can be discriminated is 20 to the power of 6, or 64 million. A learning device making no use of continuity would need to treat each of these states separately. To do this it would require a large amount of information storage, though not enough to be a problem using modern computer technology. A more serious difficulty is the time that would be needed to learn a good control policy, as this would only be achieved once every discriminable state that could possibly occur had been experienced sufficiently often that statistical data justifying an appropriate response had been accumulated. For any nontrivial control task, the learning or self-optimisation of such a device must be painfully slow. (Because inputs can include time derivatives as well as pseudo-integrated and time-displaced versions of the transducer outputs, the suggestion of six inputs is modest.)

The complexity of the learning task is of course much reduced if the input signals to the controller are subdivided more crudely than into the 20 intervals supposed above. In some situations this has proved to be acceptable, as in the study under the name of BOXES, by Michie and Chambers (1967) discussed in Chapter 2, where the situation at any instant was represented by a choice from 162 "boxes".

The crude subdivision of the four-dimensional phase space proved to be sufficient to allow the system to "learn" to balance the pendulum. That it was so was only discovered by trial, and if it had been insufficient some different and probably finer subdivision would have been tried instead. The overall learning system consisting of (automatic system plus experimenter) therefore had access to the raw continuous variables. Another point is that the problem was formed as what is termed in the AI literature a "well-defined problem", namely one for which there is an algorithmic means of checking whether any purported solution is valid. Where the requirement is to find a control policy that balances the pole indefinitely the problem is essentially well-defined (even though the balancing can only be observed for a finite time). The acquisition of manipulative skill by a person, on the other hand, does not usually stop when a particular criterion is reached. In the particular example of pole-balancing, there would be ongoing refinement to let the pole be maintained with less "wobble". The BOXES experiment on pole-balancing does not justify disregard of continuous variables as such.

It is rather usual to consider the acquisition of a manipulative skill as something distinct from learning involving discrete concepts. Skill acquisition does not usually involve the formation of a detailed model accessible to consciousness. However, the control policy that is successively refined is a reflection of the structure and parameters of the environment in which the skill is exercised, and there is no firm distinction between this and a model, as discussed further in Chapter 4. The reflection constitutes a theory, in the sense in which the term is defined in Chapter 1, because it is equivalent to a set of rules relating the response of the controlled system to its inputs. The essential feature of a theory is that its dictates agree with observations, and the process of skill acquisition operates to improve the correspondence within a certain range of situations. The bicycle rider refers to his or her internal model, perhaps consciously at an early stage of learning but subconsciously later on, and makes use of it to decide that, for example, the response that will correct a certain amount of tilt is to steer to the same side by an appropriate amount.

The acquisition of manipulative skill is a form of learning in which the criteria of similarity required by the *basic learning heuristic* are fairly unambiguous and directly related to physical variables. The task environments usually considered in AI involve discrete concepts and, as Minsky and Selfridge observe, the criteria of similarity required by the heuristic are more subtle. Nevertheless the two types of task environment have features in common, and it would be surprising if the neural means of dealing with each had evolved independently (though any suggestion about evolutionary origins must be speculative).

It is rather usual to treat the two types of task environment as distinct, and in much work in the AI context, having a bearing on robotics, the manipulative aspect is disregarded, as in the famous "Blocks World" study by Winograd (1973). That such partitioning is inappropriate when considering human performance from an evolutionary viewpoint is argued by Churchland (1986):

> There is an assumption, popular among philosophers, that the brain processes that make for cognition are one sort of thing and that the brain processes that contribute to motor control belong to an entirely different category. Accordingly, the assumption may be used to justify a disregard of research on motor control by those doing research on cognition. But if we look at matters from an evolutionary and biological point of view, the assumption is not only theoretically naïve, it in fact trammels the theoretical imagination. [The slightly unusual word "trammel" refers originally to an arrangement of nets used to guide fish to a region where they could be caught, but has come to mean "to put restraint upon, hamper, impede, confine".]

Churchland defends her viewpoint by arguments based on an imaginary robot, and it seems reasonable to expect the development of advanced robotics to indicate a probable course of human evolution in this respect. It is probably true that current work on robotics does not give much support to the viewpoint, but then the state of the art is very far from producing anything like a true android.

That robotics, and its associated problems of perception, require information processing of a kind that is unfamiliar to workers in traditional AI is acknowledged by Brady (1981) in a robotics context:

> Many observations about the world, as well as our assumptions about it, are naturally articulated in terms of 'smoothness' of some appropriate quantity. This intuitive idea is made mathematically precise in a number of ways . . .

Whether or not the treatment in this book represents the most fruitful approach, it is a strong "hunch" of the current author that an essential key to progress in AI is appropriate attention to the interface between continuous and concept-based processing. The ease with which living systems cross the interface invites description by the trendy term "seamless". Many games that people enjoy playing involve this apparently seamless mixing of strategic planning, which can be regarded as concept-based, with dexterity or manipulative skill of a continuous nature. This is for example true of competitive ball games, whether played on a field or on a table-tennis or billiard table, and also holds for many computer games. It is as though people enjoyed demonstrating, and honing by practice, this special ability not so far remotely approached by robots.

The term "concept" is intended rather liberally and need not refer to anything that is named in a spoken or written language. It can be any discrete event or development in the thought process. In game playing, it can range from a simple matter of appreciation that the referee has blown his whistle to the complex realisation that an opportunity has opened for some profitable move such as kicking the ball towards the opposing goal, or passing it to a teammate who may be able to do so. Appreciation of the opportunity depends on both conceptual and continuous processing and even on estimates of psychological factors such as the alertness of the teammate and of opponent players in the vicinity.

The type of continuity that is effective does not correspond to the mathematical definition of continuity "in the small", defined by saying that $f(x)$ is continuous at α if its limit, as $x \to \alpha$, is $f(\alpha)$. Continuity so defined is meaningless in the real environment, and the term has to be understood more robustly, as the property that allows interpolation, extrapolation and smoothing, and ranking.

There is a connection here with the early work of Gordon Pask (1962). He maintained that a process should be described as "learning" only if it involves a change of description language. Processes producing only changes in parameters he would describe as "adaptation". "Language" is not to be understood as necessarily a spoken or written "natural language" but as communication in a general sense, for instance within an individual's nervous system. Obvious examples of change of language are associated with pattern recognition, where, for example, the representation before learning might be a jumble of sensory inputs, including visual, auditory, and perhaps smell components, and after learning might constitute recognition of, say, "sabre-tooth tiger approaching".

The distinction made by Pask was more clear in the discussion after his 1962 presentation than in the printed version. I can remember Arthur Samuel, originator of possibly the best known computer learning algorithm (Samuel 1963), nodding thoughtfully as he considered the implications. His method was applied to the game of checkers (draughts) and its basis was later adapted to the classic challenge of computer chess.

It is interesting that Samuel's program depends on the conversion of a playing position on the board to a set of numerical values representing the strength of a player's position. These have names like "piece advantage", "centre control", and "mobility", which make sense to game enthusiasts, and which can be processed to allow decisions about moves to be made. The conversion is a clear example of a change of description language, from one of a definitely discrete or concept-based kind to one exhibiting continuity. However, the conversion was not performed spontaneously by the computer program but was due to Dr Samuel's own insight. The program has the means of adjusting the relative weights, or proportions in which the measures are combined, and of making trials with certain combinations, but not of producing fundamentally new criteria. Even so, it indicates the usefulness of combining continuous and concept-based representations.

The contention here is that concept-based intellectual processes have evolved from the continuous kind. This can be regarded as a change of language, in the sense intended by Pask, though occurring in evolution rather than in the learning of an individual. Another point, also discussed by Pask among others, is that the acquisition, retention, and proliferation of concepts parallels biological evolution, in that concepts are retained, and help to form further concepts, if they prove useful, but expire otherwise. The transition to concept-based operation can itself be viewed as a concept that is retained, as epitomised (Andrew 1981) by reference to "the concept of a concept".

References to continuity may be to simple metrical continuity, as in bicycle-riding, or to less tangible manifestations as referred to in connection with the *basic learning heuristic*, as applied both to learning and to evolution. The latter is similar to Minsky's *heuristic connection* between problems. Evolutionary progress depends critically on the nature of the similarity criteria that are effective, especially the criterion of similarity of solution. For evolution, a "solution" is a state of the evolving entity and "similarity" is the readiness with which one state can convert to another.

If a computer program is to be successively improved in its performance of some task, changes have to be made in a functional representation. The alternative of changing a few characters in a program listing (in "machine language" or in a procedural language like Pascal) is not likely to be successful, especially as most changes will cause the "program" to "crash", or to cycle indefinitely, or to fail to compile. It can be said that in that case the choice of representation and hence of similarity of solution mean that mutations are almost always fatal. An early study by Friedberg (1958) was aimed at automatic improvement of a trivially simple computer program and had very limited success, as discussed by Minsky (1963).

The choice of representation, and of similarity criterion within it, must be such that moves between similar states are only sometimes disastrous. This could be achieved by making the criterion conservative, for example by only recognising states as similar when they differ only in simple respects like settings of numerical parameters. On the other hand, an unduly conservative criterion must correspondingly limit the evolutionary possibilities. Biological evolution must have resulted from a particularly fortunate choice, or sequence of choices, of representations and criteria.

The difficulty of choosing suitable criteria is illustrated by the comments of Oliver Selfridge (1959) on his "pandemonium" scheme. The name is chosen to emphasise the assumption of parallel processing. In the scheme a set of "computational demons" evaluate respective features of presented patterns, and their outputs are used as a basis for pattern classification. The signals go, first, to a set of "cognitive demons" corresponding to pattern types to be identified and then to a "decision demon" that selects one pattern type.

The details of the classification method are not important for the current argument. The important point is that each computational demon receives feedback of "worth", a quantity designed to indicate how useful that demon's contribution to the classification process has proved to be. The feedback of "worth" can be used to modify the network as though by biological evolution. Computational demons that have low "worth" are eliminated, and new ones are formed to be similar, but not identical, to those that proved to have high values.

Two methods are suggested for creation of new units, one termed "conjugation" and the other "mutated fission". The former was implemented without difficulty because it is simply the combination of the outputs of two preexisting high-scoring demons, using a randomly chosen form of combination. The more interesting "mutated fission" requires the formation of a new demon that is similar to a high-scoring one, and here the choice of an appropriate representation of the demon, with its associated criteria of similarity, is critical. Selfridge appreciates the challenge and difficulty of this (where the preexisting demon is referred to as a "subdemon"):

> This will usually require that we reduce the subdemon himself to some canonical form so that the random changes we insert do not have the effect of rewriting the program so that it will not run or so that it will enter a closed loop without any hope of getting out of it.

The canonical form must balance the conflicting requirements of robustness with regard to the pitfalls mentioned against flexibility to allow versatile evolution. Its choice is clearly extremely critical, and it is significant that Selfridge does not offer a concrete example. The difficulty is similar to that of automating the devising of qualitatively new useful criteria for the checkers-playing program.

Near Misses

A feature of learning, and possibly of evolution, that tends to be overlooked, is the need for attention to what can be termed "near misses". It is important in the current context because it has continuous aspects. When a person learns a potentially hazardous skill, such as driving a car, the first stage of training is probably closely monitored by an instructor, but at some point the learner is on his or her own and learning by direct interaction. Other potentially hazardous skills may be acquired by direct interaction from the start. The fact that this can happen with a tolerable survival rate shows that learners take account of "near misses" and adjust their policy accordingly.

Sometimes appreciation of a "near miss", or avoidable dangerous situation, will depend on careful analysis and exercise of imagination. In many cases, though, the lesson to be learned will be simply that some tricky part of a journey should have been taken more slowly, for well-known reasons connected with vehicle stability and with reaction times of the driver and other road users. The process of learning from the "near miss" situation then has continuous character.

An interesting aspect of the car-driving situation is that people seem to converge on a particular level of danger. Safety is not the only consideration and people like to go fast, both to save journey time and as a remarkably strongly felt matter of personal prestige. The result is that technical advances that might be expected to increase road safety are often compromised by driver behaviour, so that, for example, improvements in braking and road-holding do not improve safety as much as they might because people start to drive faster and closer. This is again a form of learning having continuous character.

A different example of "near miss" learning, associated with a more primitive stage of evolution, is the testing of plants and other substances for edibility. This could obviously prove fatal but may be survived if only a small amount is eaten. The stomach then has the means of giving a warning of the "near miss".

Logic

The word "logic" is usually understood to imply the manipulation of discrete concepts, as in "computer logic" and to differ in this respect from natural language. The use of "natural language" and its refined form as "logic", both for interpersonal communication and for internal reasoning, is the main thing that distinguishes humans from other animals. Humans are not completely alone in using language, and elaborate communication occurs between members of other species, such as honeybees. Also, some animals, notably chimpanzees, have learned something of human language.

Nevertheless, no other species can boast of anything remotely approaching the magnificent and useful structure that is "natural language". Perhaps arising from collective vanity of the race, there is a rather common assumption that language is the basis of all thought. The refinement to "logic" is often assumed to give still greater power. That language is not the sole basis of thought seems fairly obvious from introspection, as ideas and plans and impressions come into being in ways that cannot be accounted for by purely verbal processing, even if many of them are promptly given verbal expression. The very considerable intelligence of many animals confirms that use of an advanced language is not a necessary precondition.

The use of analogy in argument usually depends on a correspondence between the sides of the analogy that would be difficult to defend or even explain linguistically. The use of heuristics in AI has a similar character because a heuristic originates as a guess by the experimenter, to be retained or discarded depending on its effectiveness in serving some purpose.

That spoken or written language represents merely the "tip of the iceberg" of the thought process, and that continuous processing underlies it, was argued in Chapter 2. The importance of continuity can also be argued from other, rather more obvious, linguistic considerations. The usual interpretation of "logic" is as a form of processing that rejects vagueness and half-measures. The world is seen in black and white. On the other hand, the real world is not black-and-white, and even these terms have shades of meaning within them.

The fuzzy methodology introduced by Zadeh provides one way of dealing with graded characteristics, by reference to *linguistic variables*. An example would be "tallness" in reference to "a tall man". Fuzzy techniques are undeniably useful in combining features of continuous and discrete operation, but there is reason to doubt whether they combine them in the same way as the brain does. It would seem to be more realistic to regard linguistic variables as useful in their own right and to extend theories about concept-formation to refer to them as well as to discrete concepts.

The term "linguistic variable" is misleading, because the origins of intelligence were presumably prior to the emergence of spoken language. A better term would be "quantifiable concept". Kochen (1981) has argued for the effective utilisation of concepts such as "edibility" at an evolutionary stage where they could only be represented in the "language" of the nervous (or endocrine, etc.) system of the animal. Of course, the attribution of a "concept" in a nonverbal context has to be a descriptive expedient adopted by the observer, but one that is compelling.

The importance of continuous variables is demonstrated by characteristics of the "natural language" that serves everyday needs. This uses adjectives and adverbs, and these are expected, as standard, to have comparative and superlative forms, implying that they refer to continuous characteristics of the associated objects or actions. It is taught that certain adjectives, such as "perfect" or "unique", should not be qualified because their meanings are absolute – something is either unique or it is not. Even here, though, everyday usage and the usual perception of the meaning suggest degrees of uniqueness, sometimes captured by some such phrase as "outstandingly unique".

It is sometimes suggested, especially in science fiction, that a super-powerful extraterrestrial intelligence can operate by not being, as Partridge (1986) puts it, "impeded by human weaknesses such as emotions and non-logical reasoning." He refers in particular to the character "Spock", whose portrayal in the early runs of the television series *Star Trek* he refers to as the "Spock myth". As *Star Trek* fans (such as myself) will know, the plots of many of the episodes were such that Spock was obliged to expand his outlook.

AI and Computers

AI workers tend to accept discrete logic as the basis of their efforts, though as a great variety of approaches come under the heading, it cannot be said that continuity is entirely ignored. As has been noted, the learning algorithm due to Samuel (1963) has the interesting feature that a discrete situation is converted to a set of continuous descriptors. This is specially interesting in connection with the Ashby–Bellman debate that will be mentioned later.

In many situations, discrete indicators arise on the basis of continuous variables. This is particularly clear when recognition of something like printed characters is involved. This is likely to depend on a continuous measure of correspondence between a presented character and members of a set of exemplars, out of which one is chosen as best match. The continuous measure could be spatial cross-correlation of the images after adjustment to agree in position and perhaps size and orientation.

The adoption of discrete logic as the essence of intelligence was enshrined by the work of McCulloch and Pitts (1943) when they used the possibility of such processing as their criterion of the capability of neural nets. In fact, real neurons are usually observed acting as continuous rather than discrete "logical" elements. The emphasis on discrete logic is sometimes attributed to the emergence of digital computers, although they too have associations with continuity to a greater degree than is often supposed.

Rather curiously, the rise of AI, with its emphasis on discrete logic, was associated with an emphasis on continuous processing and spatial visualisation, an opposite trend. The now-ubiquitous computer mouse was invented in 1963 and was enthusiastically adopted as a means of man–machine interaction by the AI fraternity, with the argument that it felt natural to a human user, a clear acknowledgement of human preference for continuous operation. At the time, most computer users expected to use punched cards or paper tape as computer input and output, and the move to graphical and mouse-driven methods showed remarkable foresight. (I can remember, at the time, feeling sure the idea would not catch on!)

The computer mouse was not the only acknowledgement of the value of continuous processing. I can remember visiting MIT in the late 1960s, when my experience of computers was of the tedious punched-cards-and-tape kind, and being shown by Marvin Minsky a computer with a display screen and keyboard, running a program that allowed him to tap in a mathematical function of x and promptly to see on the screen a graph representing it over a range of values.

The computer had also been fitted with continuous input devices of the joystick variety which allowed a game of "space war", in which each participant controlled a space ship and could fire simulated missiles at the other's ship, with also the possibility of shooting down enemy missiles in midflight. My recollection is that my performance in the game, as a complete beginner, was not too bad when the play was without simulated gravity, but when this was introduced, giving ships and missiles acceleration towards a central "sun", for me the situation was hopeless. Marvin, however, was able to operate under these conditions, due to much practice.

What we were playing with was of course a forerunner of the computer games that are now commonplace and as remarked earlier are similar to many traditional sports in their intimate mixing of continuous and discrete skills. A similar mixing is a feature of the other kinds of continuous-variable interaction. The display of the graph of a mathematical function required, first of all, parsing of the algebraic form to derive "machine language" rules for evaluation, followed by the continuous-variable process of successive evaluations. (Strictly, if the program was in an interpretive language, these two components may not have been so distinct as this suggests, but both would be present in essence.)

The ways in which the computer mouse is used in practice also involves intimate mixing. In most cases its continuous input is promptly converted to a discrete form when a choice is made from a menu or an array of icons, or a word or insertion point is chosen in a piece of text. The term "WIMP" has been used to describe the usual modern computing environment, where the acronym stands for "windows, icons, mouse, and pull-down menus". The WIMP environment only became feasible once computing power was seen as cheap enough that it could be expended in repeatedly interrogating input devices such as the mouse and keyboard, and sometimes also joystick-type devices needed for computer games. The importance attached to the mouse and these other devices confirms that continuity is believed to be a part of "natural" interaction.

The common assumption that digital computers are fundamentally associated with binary arithmetic and two-valued logic is misleading. It is true that they can be built using only the binary-logic "gates" (AND, OR, and NOT, or other combinations) that can also be realised as McCulloch–Pitts model neurons. However, the prime motivation of the pioneering work of Babbage was numerical computation, and for this the adoption (subsequent to Babbage) of binary arithmetic is essentially a matter of economics. The techniques available for building computer hardware are such that reliable storage and processing are more readily achieved using binary arithmetic than, say, decimal, even though the number of digits needed for a given precision is greater by a factor of just over three. For manual computation the considerations are different, as people do not have great difficulty in remembering the addition and multiplication rules for decimal arithmetic and prefer decimal notation for its relative compactness.

Although the work of Babbage was directed towards numerical computation (especially the compilation of tide tables, where direct coupling of the computer to the printing process was seen as a valuable feature, eliminating human transcription errors), he and his collaborator Ada Lovelace undoubtedly visualised much more, with even suggestions that machines would manipulate symbols as well as crunch numbers (Bowden 1953; Dubbey 1978). An absolutely critical development, which they appreciated and hoped to include in a computer based on the Jacquard loom, was the means of letting the future course of a computation depend on results already obtained. In high-level programming languages, the result is usually produced using a statement embodying "if" followed by "then" and perhaps "else", or in a slightly less general form within the rules determining the number of repetitions of program loops.

It was this development, rather than an association with formal logic, that paved the way for the vast burgeoning of computing capabilities. In the context of the discussion here, it is interesting that it allows a connection between continuous processing and the discrete choice inherent in the "if ... then" statement. (The choice is usually binary, at each stage of presenting "if" followed by a condition, but this is a matter of convenience. Some languages, including ALGOL68 with its "case" construction, allow choice from more than two possibilities.)

The data transformations carried out by the CPU or "mill" of a computer can be of either a "logical" or "arithmetic" nature. In the latter the contents of computer "words", or concatenations of them, are treated as numbers, so that, for example, in addition or subtraction there are carried from place to place as in pencil-and-paper arithmetic. When operating arithmetically, with suitable digital-to-analog and perhaps analog-to-digital transducers for output and input, a digital computer has similar capability to an analog one. The digital-to-analog conversion is likely to be included in the arrangements for display on a screen or other graphical device. With appropriate transduction, or if the user is prepared to do manual plotting from a mass of numerical output, a digital computer can deal with data that is continuous in the sense used here, and analog machines are now seldom if ever used. The output of the digital computer is necessarily discontinuous at some level, but the "digitisation noise" can be very small.

There was of course a time when the means of interacting with digital computers were more limited, and also their internal working was less suitable for this type of task, especially in not having floating-point number representation readily available. For this reason, MacKay's defence of continuous processing (MacKay 1959) is partly expressed as a defence of analog computing methods, especially in an earlier paper of 1949 included as an appendix to the 1959 presentation. At the time he was writing, analog methods offered the possibility of altering parameters, perhaps in a simulation of a dynamic system, simply by turning knobs, and of seeing the result practically instantaneously. Digital methods did not then offer such convenience, and he gives reasons for believing they were unlikely to do so, though he admits the possibility.

When the details of its working are considered, the use of floating-point arithmetic seems to be a very cumbersome way of achieving a simple result of addition or multiplication in, say, a simulation of a dynamic system, and often the computation is performed with much greater precision than is needed. Part of MacKay's argument was that this unnecessary accuracy must be paid for in reduced speed of operation. This is true up to a point, but since his time of writing solid-state electronics has come into being, with enormous gains in speed and reliability, and a great deal of attention has been given to the design of processors for fast operation. Cumbersome though its details may seem when described, fast and reliable floating-point operation is readily available and is an expected feature of an installation. Continuity is an important feature of digital computing as well as of the nervous system. (The computing power of supercomputers is sometimes expressed in gigaflops, or units of 10^9 floating-point operations per second, with even a projected petaflops machine, capable of 10^{12}.)

The use of this technology is of course visualised in situations where the computational task has been specified by a programmer. The internal working of the processor has to be inviolate, because any small alteration will almost certainly result in complete failure. In this respect the operation of the digital processor differs from that of an entirely analog component carrying out an arithmetic operation, where a small change in the physical characteristics of the component probably produces a correspondingly small change in its operation. For example, a component forming a weighted sum might come to use a different set of weights, and characteristic curves of nonlinear operations could come to have different shapes. In other words, analog devices are likely to have measures of similarity of structure and of effect that allow evolution or learning at the component level. (The forming of a weighted sum is a feature of the elements of most schemes for artificial neural nets.)

Stafford Beer (1959) stresses this evolutionary aspect in an appendix entitled "The analogue of fabric" with a claim that the digital computer is unsuited to be a model of an adaptive viable system such as a firm, and that some kind of organic model is more appropriate, though the kinds of model that he tentatively suggests have not proved useful except as illustrations.

Stafford Beer was of course writing when digital computers were much less powerful than they are now, and since then many programs have been written that usefully simulate biological evolution, either for its study as such or for exploitation of its mechanisms in other contexts. Despite the inflexible character of the digital computer at the component level, effective randomness can be introduced using a special routine termed a pseudo-random number generator.

There is however a sense in which Beer is right, because the evolutionary changes that are possible in the digital simulation must to some extent have been allowed for by the programmer, even though it is usually only by creating and running the program that the full implications can be determined. The alternative of using some organic simulation also has disadvantages because, although adaptive viable systems can be argued to have properties in common, it is impossible to say what these are, except superficially, and the relationship between the system of primary interest and its organic simulation must be uncertain.

In all the discussions of adaptive systems, the considerations of continuity referred to in the *basic learning heuristic* are implicit. The observation that computers do not allow adaptive changes at the component level has an obvious correspondence to the unsuccessful attempt of Friedberg (1958) to achieve automatic optimisation of a computer program.

The Ashby–Bellman Debate

A debate having a bearing on these matters, very early in the history of AI, was between Ross Ashby (1964) and Richard Bellman (1964) at a meeting in Evanston, Illinois. Ashby gave a keynote address, in which he suggested that a good way to stimulate progress in the following ten years would be to promote a major effort to make a chess-playing machine that would beat the human world champion.

Bellman contested this view on the grounds that chess operates in a highly artificial environment quite different from that in which natural intelligence has evolved. In particular, the chess environment lacks the continuity that allows convergence on a solution. Most tasks in the everyday environment have a different character. The discovery of a procedure for baking a palatable loaf of bread, for example, was undoubtedly achieved by producing a succession of loaves that were suboptimal though not totally unacceptable.

There are senses in which both Ashby and Bellman were right. Chess is undoubtedly highly artificial and not the best context for examination of the origins of intelligence. On the other hand, people undoubtedly do learn to play chess and it seems safe to assume that, somehow, the *basic learning heuristic* becomes effective. (The occasional occurrence of prodigies who take to chess as though by instinct is a puzzling phenomenon for which no explanation is offered here.) The approach to game-playing due to Samuel (1963) indicates one way that a discrete board configuration can be represented by continuous measures, and it seems likely that human play makes use of such transformations. There is still much to explain, however, as even in the simpler environment of checkers (draughts), Samuel's refinement of means of combining the respective measures depended on a great deal of exploration of the tree of possible game continuations, an activity that does not seem to be performed by human players, at least not consciously.

As mentioned in Chapter 2, chess-playing programs depend heavily on exploration of the look-ahead tree, though with economy due to employment of the *alpha-beta technique*. This delivers results that can be shown to be equivalent to those obtained by exhaustive development of the tree to a given depth, but with large parts of the tree ignored. Techniques are fully treated in the AI literature, particularly Nilsson (1971).

Continuity also enters into traditional AI studies in another way, having some correspondence to MacKay's reference to weighting of conflicting bodies of evidence. Minsky (1959a) refers to a class of heuristics termed "administrative methods", of which some come under one of the three headings of (a) methods that set up new problems as subgoals, (b) methods that set up new problems as models, and (c) methods of characterisation or description. The General Problem Solver of Newell, Shaw, and Simon (1959) is a clear application of methods under heading (a), whereas both (b) and (c) depend on the postulated *heuristic connection* between problems. The importance of continuous measures is demonstrated further by the observation that to employ the methods effectively, it is necessary to form estimates of the difficulty, or computational cost, of subproblems or model problems, and of their probable utility when embarked upon.

A related idea is a measure of "interestingness" or "worth" of a result, as part of a scheme discussed by Lenat (1984). (Measures of "worth" are also a feature of Selfridge's *pandemonium* scheme, discussed at length in Chapter 5.) Other examples are measures of "salience" and "prototypicality" introduced by Hall (1989) in a comprehensive review of approaches to analogical reasoning. "Salience" represents the suitability of a source structure for the forming of analogies, and "prototypical-ity" refers to the corresponding property of a target structure, namely its readiness to be usefully extended by analogy.

Presumably a criterion of "interestingness" accounts for the attention given to some theoretical topics that apparently do not fit with the discussion of theories in Chapter 1, where a theory was judged by its explanatory power relative to observa-tions, in an essentially utilitarian context. Pure mathematics and some other topics appear to have no justification in these terms and yet are pursued with enthusiasm. This is presumably because their complex self-consistent bodies of theory score highly in "interestingness" as judged by criteria formed in other contexts. Accord-ing to this view, "pure" mathematics has a stronger connection with practicality than some of its adherents would like to believe, and it is hardly surprising that many findings initially thought to be "pure" eventually find applications. The per-ceived need to introduce these various considerations supports the contention that continuous measures are important aspects of intellectual activity.

Summary of Chapter 3

The importance of continuity as an aspect of intellect is defended on various grounds, including the nature of the perceived environment and reference to a num-ber of continuous variables that arguably play an important part in thought pro-cesses. These include estimates of distance from a solution in problem solving, and weightings attached to items of conflicting evidence. Continuity is implicit in Ashby's graphic representations of state-determined systems.

Contrary to the apparent implication of the neuronal models of McCulloch and Pitts, continuity is apparent in nervous activity. It is indicated by the nature of recorded neural activity and by functional considerations, the latter including perfor-mance of acts of skill and participation of neural action in regulation of such things as body temperature and amount of light entering the eye. Discrete responses are also formed, and it is argued that this is likely to be due to positive feedback in pri-marily continuous pathways, in a manner similar to the achievement of "flip-flops" in electronics.

Continuity has to be an important part of any process of evolution or learning. Whatever structure is modified by the process must have criteria of similarity of its states, and criteria must exist also for the environment in which it operates. The two criteria of similarity are sufficient to permit evolution or learning, but it will be more efficient if a continuous mapping exists.

The use of continuous criteria in adaptive process control and in game playing, introduced in the previous chapter, are discussed further in this context. It is observed that people combine concept-based and continuous processing in a remarkably "seam-less" fashion and seem to enjoy demonstrating this in games. An extra feature not usually invoked in this context is appreciation of the need to pay attention to nonfatal occurrences of the kind that would be "near misses" in learning to drive.

Logic and language are discussed, with the observation that everyday needs are best served by natural language in which comparative and superlative forms of adjectives and adverbs are standard, further supporting the value of continuity. It is also argued that digital computers are less firmly rooted in discrete logic and binary representation than is often supposed, and, like neurons, readily deal with signals that are effectively continuous.

Reference is made to a historical debate between Ross Ashby and Richard Bellman in which important considerations were aired. Some further uses of continuous measures in thought processes are reviewed, particularly administrative methods in problem solving as discussed by Minsky.

Chapter 4
Adaptation, Self-Organisation, Learning

Adaptation in Continuous Environments

Adaptation in continuous environments has received much attention under the heading of control engineering, with one class of solution known as the Kalman filter (Sorensen 1985). Other schemes already mentioned include Gabor's "learning filter" (Gabor, Wilby, and Woodcock 1961) and similar proposals by Andrew (1959b) and the adjustments embodied in Samuel's checker-playing program (Samuel 1963).

The latter group of schemes has the simplifying feature, as far as discussion is concerned, that no account need be taken of dynamics, though dynamic control can be achieved by letting some of the inputs be derivatives or pseudo-integrated or time-displaced values. In the description of the BOXES method applied to balancing an inverted pendulum (usually termed "pole-balancing"), two of the inputs were derivatives.

An additional difficulty in the dynamic situation is in choosing an appropriate indication of hedony, since the results of an input appear after some delay, and the appropriate measure may be a value integrated over a period. This does not arise in the BOXES example where changes in parameters are made after a run, with the measure of hedony the time to failure, which is when either the pole falls or the trolley leaves the track. If the system was required to improve further and to balance the pole with less "wobble" after continuous balance had been achieved, a different criterion would be needed.

Similar techniques have been studied under the heading of "reinforcement learning", where of course the feedback of hedony can be termed reinforcement. Sutton and Barto (1998), and contributors to Sutton (1996) have added extra features that greatly enhance the operation in many contexts. These workers consider the learning device in finite-automaton terms, such that an action not only brings an immediate (or at least, short-term) reward but also causes a change of state. The long-term return from an action is therefore the sum of the immediate return and that expected to be achievable from the new state. The choice of optimal action therefore depends on estimates of value associated with possible successor states, intended to represent the mean return achievable from each.

A.M. Andrew, *A Missing Link in Cybernetics*, IFSR International Series on Systems Science and Engineering 26, DOI 10.1007/978-0-387-75164-1_4,
© Springer Science+Business Media, LLC 2009

The principle is similar to that of "dynamic programming" (Bellman 1961; Dreyfus 1961). The estimates of value associated with the respective states are modified as the system operates, as a part of the learning process. There is a strong connection with Samuel's method of "bootstrapping" values of board configurations in his checker-playing program. The states may be discreet or may form a continuum, and with the latter alternative the method has been employed in robotics. Sutton and Barto demonstrate the power of the method with various examples including a very highly successful self-improving program for backgammon. However, neither this enhancement of the general principle, nor the consideration of response time mentioned earlier, is embodied in the current treatment.

As suggested in Chapter 2, the control action d is assumed to be computed as a function of the transducer signals a, b, c, and so on, perhaps as a weighted sum of these and higher-order terms

$$d = K + La + Mb + Nc + \text{higher-order terms}$$

and adaptation consists of the adjustment of the parameters K, L, M, and so on, and possibly also selection of terms to be included, so as to maximise some indication of success, represented by a numerical value h. The letter is suggested by the attractive term "hedony" used by Oliver Selfridge to refer to such a quantity.

In many situations the evaluation of h is not a simple matter, but here for simplicity many of the difficulties will be ignored. Where a dynamic system is controlled, a useful evaluation of h is likely to be possible only after a delay. In the pole-balancing experiments, the indicator of success was the length of time after a decision before the pole fell over (or the trolley ran off the track), so became available after a variable delay. In the application of BOXES to learning to play noughts-and-crosses (tic-tac-toe), not described here, the indication only became available at the end of each game. For more complex games a different method is needed, as will be described.

As noted earlier, one of the ways in which the pole-balancing experiments failed to model human performance was in ceasing to make further adjustments once the pendulum was kept up for a time sufficiently long that it could be assumed it would remain so indefinitely. This probably led to a dynamic performance that was amusing to watch, but a person usually carries the learning process further so that the pole is kept more and more steady. The useful indication of h is then obtained by integration over an interval that depends on the dynamics of the controlled system, and an early stage of the human learning process is presumably to form an estimate of the system's response time as far as the estimate of h is concerned, an estimate that is probably revised as skill is acquired.

Adaptation in an unknown continuous environment has to include associated optimisations of both this response time of the controlled system and of a suitable time-constant for smoothing of the statistical estimates used. What is discussed as simple optimisation must actually be achieved by the interaction of at least three processes of optimisation. This complexity is likely to be obscured in experimental studies because various parameters including the time characteristics are usually adjusted or "frigged" by the experimenter until the scheme works well. The experimenter himself provides feedback, perhaps unwittingly, that is not available externally to an autonomous organism.

For living systems, it seem safe to assume that adaptation is with regard to survival. It is easy to see that this goal has been broken down by evolution into a great many subgoals to which hedony measures can be ascribed, as numbers or at least as rankings, associated with nutrition, temperature, reproduction, and so on, but for the present this hierarchy will be ignored because the focus is on evolutionary origins. Wiener (1948) used the term "affective tone" equivalently to "hedony".

Considerations relevant to real-world or online adaptation correspond closely to methods of optimisation of computer models, admirably reviewed by Schwefel (1981). Optimisation is often subject to constraints, and in problems lending themselves to solution by methods under the heading of *linear programming*, constraints form the main part of the problem statement. Constraints are another feature that will be ignored here for simplicity. There is justification for this in the observation that constraints can be incorporated in the means of evaluating h, which corresponds to the *objective function* in computer optimisation. The algorithms used computationally are such that it is usually advantageous to evaluate the objective function without regard to constraints and to introduce them separately, but it is easy to feel that biological adaptation probably differs in this respect.

Even and Odd Objective Functions

Tasks of adaptation or optimisation, or of learning in a continuous environment, can have either of two very different characters, and the distinction has sometimes been surprisingly ignored. In one situation, some trial output (values of a, b, c, etc.) from the adaptive controller produces a value of h, but without any indication of how these output values could have differed to produce a more favourable outcome. This is the pure "hill-climbing" situation (with the assumption that the climber is blind apart from being aware of altitude and of his own latitude and longitude). It can be said there is an even objective function because, if the top of the hill is rounded, the altitude falls away on both sides of the optimum value of a variable, and so is predominantly an even function of that variable with the optimum as origin.

There are many situations, however, where the evaluation of h is associated with an indication of how the response of the controller might have differed to give a more favourable result. This is so in any situation where it is appropriate to refer to an error in the control, and so to "learning with a teacher". Any situation where the task is to form a model, or to make predictions, has this character because a "correct answer" becomes available for comparison. The optimal result is zero error, with the error indication taking different signs, with respect to a given variable, on opposite sides of the zero-crossing taken as origin. It therefore has the character of an odd objective function.

Much of human learning combines elements of both the "with a teacher" and the hill-climbing types. For many tasks, learning is partly by imitation or by instruction, and so classed as "with a teacher", but subsequent development of the acquired skill occurs without such guidance. Also, the phase of learning by imitation or instruction will usually build on basic skills of, say, bodily posture and balance, and manipulation of objects, that were acquired and honed previously. However, for the purpose of the current analysis, and certainly for consideration of the various attempts to make computer models of aspects of learning, the two types will be considered separately.

Trainable pattern classification schemes of the *perceptron* kind are usually described with reference to just one response unit, which is meant to respond in one way when a presented pattern belongs to a particular class, for instance the letter A, and to respond differently to patterns not belonging. It is assumed there will be a set of response units, one corresponding to each pattern class to be recognised, the pattern classes perhaps corresponding to the letters of the alphabet. For each response unit the feedback from a "teacher" is an indication of right or wrong, and clearly where there are only two possibilities, an indication of "wrong" is also an indication of what would have been "right" for that unit, placing this in the odd-objective-function category.

The means of adjusting weights in Samuel's checker-playing program also depends on feedback of the odd-objective-function kind, although it is not strictly error information. Use is made of the fact that a particular static evaluation function can be used in either of two ways to select the next move to be made in the game. The simplest way is to apply evaluation to the configuration that would result from each possible move and to select the move that scores highest. A better, but computationally more expensive, method is to form the tree structure representing all possible game continuations (or a subset chosen by some criterion of plausibility) to a certain depth and to apply the evaluation function to the positions represented at terminal nodes or "leaves" of the tree. The values obtained are then fed back through the tree by a "minimax" procedure (choosing the maximum when the move is by the program and the minimum when it is by the opponent) to produce evaluations of the possible moves and from these to make a choice as before. (The static evaluation function is so called because it does not involve any lookahead.)

It has been found that the quality of play, using a given static evaluation function, increases with depth of lookahead. Samuel therefore arranged to adjust the coefficients in the evaluation function so as to improve its correspondence, when applied directly, to the result obtained by applying it at the terminal nodes and minimaxing the values back up. The direction of adjustment was therefore indicated by the method, though it was not an error as usually understood.

The distinction between true hill-climbing and adjustment using error-information has surprisingly escaped the notice of at least two eminent workers. In their "learning filter", Gabor *et al.* (1961) used a hill-climbing method although most if not all of the applications they described allowed the use of error-information. In consequence, Lubbock (1961) was able to propose a much more efficient method making use of it. Selfridge, too, in his famous "Pandemonium" paper (Selfridge 1959), suggests the use of hill-climbing in a situation where a method using error-information would be preferable.

Optimisation without a Model

There are two main ways of adjusting the multipliers K, L, M ... in the equation shown earlier

$$d = K + La + Mb + Nc + \text{higher-order terms} \qquad (4.1)$$

so as to maximize the hedony h. One is essentially model-based in its derivation and the other not, but it can be shown (Andrew 1967) that the results are similar. It is convenient to use a somewhat different notation in which the control action is denoted by c and the outputs of the transducers in the process plant as f_1, f_2, $f_3 \ldots$, with f_0 equal to unity. The multipliers altered in the course of optimisation are denoted by k_0, k_1, $k_2 \ldots$. All the quantities vary with time, the multipliers relatively slowly as part of the adaptation, but plain symbols will be used rather than the functional representations $c(t)$, $f_i(t)$, and so forth, and overlining will indicate mean values over some period, or smoothed "running" values.

The control equation becomes:

$$c = \sum_{i=0}^{n} k_i f_i \qquad (4.2)$$

where the f_i include the signals from the controlled process, previously called a, b, $c \ldots$ as well as, possibly, products and other functions of them, so allowing the higher-order terms of Eq. (4.1).

The requirement is to adjust the k-values so as to increase h. Because there is no error-information to indicate the direction of change of any value, fluctuations can be superimposed, and then positive correlation between a fluctuation and h indicates that the associated k-value should be increased, and conversely for negative correlation. Let κ_i be the superimposed fluctuation on k_i so that the equation becomes:

$$c = \sum_{i=0}^{n} (k_i + \kappa_i) f_i . \qquad (4.3)$$

The addition of separate fluctuations to all the parameters is not an elegant form of operation, however, and has the disadvantage that the overall effect on the control output c is not directly controllable. An alternative is to add one fluctuating quantity c_f to c so that the equation becomes

$$c = c_f + \sum_{i=0}^{n} k_i f_i \qquad\qquad (4.4)$$

It is then necessary to find some way of ascribing "effective fluctuations" to the k-values. If κ_i' represents the effective fluctuation ascribed to k_i, it seems sensible to set:

$$\kappa_i' = c_f / f_i \qquad\qquad (4.5)$$

According to this, the effective fluctuation applied to a k-value is that which, in it alone, would have been equivalent to c_f. The principle can also be defended in terms of the sensitivity of the output c to changes in each k-value.

However, operation cannot be exactly according to Eq. (4.5), assuming that the f_i ($i > 0$) are adjusted to have mean zero, because the effective fluctuation would become infinite at each zero-crossing. It is of course possible to let the effective fluctuation be according to Eq. (4.5) but subject to a limit on magnitude. That is to say, the value would be according to Eq. (4.5) provided its magnitude was less than some value G, but otherwise would be $+G$ or $-G$.

This plausible way of computing effective fluctuations has not proved satisfactory when tested in a computer simulation. The result could perhaps have been anticipated from the fact that κ_i takes its arithmetically greatest values when the associated f_i is close to its mean value of zero. This means that the measure of correlation of κ_i' with h receives its greatest contribution when the deviation of f_i is small, which is clearly not what is needed.

Alternatives to Eq. (4.5) that allowed effective optimisation in computer simulations were

(a) $\kappa_i' = c_f \text{sign}(f_i),$ (4.6)

where it is assumed that the f_i ($i > 0$) are adjusted to have mean zero, and sign() takes values $+1, 0$ and -1, and

(b) $\kappa_i' = c_f f_i,$ (4.7)

where again it is assumed that the f_i ($i > 0$) are adjusted to have mean zero.

Each of these performed well, with little to choose between them, in computer simulations of a simple process and controller. An interesting point that will now be demonstrated is that operation according to (b) corresponds closely to what could have been arrived at using a model of the process.

Optimisation with a Model

Given a record of operation of the process and controller over a period, a model can be formed using the statistical procedure of regression analysis. This can be formed to give a prediction of h, denoted by $\hat{h}$, as a function of the f_i and c. The standard procedure arrives at a linear function, so that $\hat{h}$ is mapped as a hyperplane in the space determined by the input variables, but if some of the inputs to the procedure are in fact multiples or other nonlinear functions of the primary inputs of c and the f_i, the result can be nonlinear in these.

The procedure requires the solution of a set of simultaneous equations, or, equivalently, inversion of a matrix. For the current purpose it is more convenient to use a variation of the method that avoids this, and there is some reason to prefer the alternative on biological grounds. In the standard procedure, where a quantity y is to be estimated as a linear function of $x_0, x_1, x_2 \ldots x_n$, we can let

$$\hat{y} = a_{0_x} x_0 + a_1 x_1 + a_2 x_{2\ldots} \ldots a_n x_n \qquad (4.8)$$

and it is usual, and convenient, to use a least-squares criterion of fit. The requirement is that the sum, or mean, of squares of discrepancies between $\hat{y}$ and y over a set of observations be minimised, where each $\hat{y}$ is computed using the values for the x_i associated with the corresponding observed y. The values of the a_i are chosen so as to minimise this, by equating derivatives to zero, leading to the so-called normal equations, of which that from differentiation with respect to a_i is

$$\sum_{j=0}^{n} a_j \overline{x_i x_j} = \overline{x_i y}. \qquad (4.9)$$

This can be written in matrix form as

$$\mathbf{X'Xa = X'Y} \qquad (4.10)$$

and solution for the a_i is equivalent to inversion of the square matrix $\mathbf{X'X}$ and simple multiplication.

If the x_i for $x > 0$ are adjusted to have mean zero and are uncorrelated, the square matrix has zero values except on its main diagonal and solution for the a_i is very simple, as:

$$a_i = \overline{x_i y} \Big/ \overline{x_i^2}. \qquad (4.11)$$

This simple expedient can be used if the raw data are preprocessed so that the x_i satisfy the orthogonal or zero-correlation condition, an expedient used by Lubbock (1961), using the Gram–Schmidt orthogonalisation method. There is some reason to think that such a method could have correspondence to biological processing, as it has been argued by Barlow (1959) that reduction of redundancy is a central feature of neural processing.

The current requirement is to form a model to predict h as a function of the f_i and c, $(0 < i \le n)$, whereas in Eq. (4.2), the f_i include the signals from the controlled process, as well as, possibly, products and other functions of them. Now, though, a larger set of such values will be used, denoted by f_i $(0 < i \le N)$ where the terms for which $n < i \le N$ are functions of c.

The orthogonalisation procedure replaces the f_i by a set of orthogonal signals θ_i computed as:

$$\theta_0 = f_0 = \text{(constant term) unity,}$$
$$\theta_1 = f_1 - a_{10}\theta_0,$$
$$\theta_2 = f_2 - a_{21}\theta_1 - a_{20}\theta_0, \tag{4.12}$$
$$\vdots$$
$$\theta_N = f_N - a_{N,N-1}\theta_{N-1} - \ldots$$

where

$$a_{ij} = \frac{\overline{f_i \theta_j}}{\overline{\theta_j^2}}. \tag{4.13}$$

The effect is to subtract from each variable, considered as a vector, the projection onto it of each previously treated variable.

The control action c as it was during the period of observation can be represented as

$$c = k_0\theta_0 + k_1\theta_1 + \ldots + k_n\theta_n + \theta_{n+1} \tag{4.14}$$

where now the k_i are not in general the same as those of Eq. (4.2) because they refer to the orthogonalised variables. It is necessary that the control action includes the term θ_{n+1} that must not be a linear combination of the other θ_i because otherwise c would be eliminated in the orthogonalisation. The extra term corresponds to c_f in Eq. (4.4). It will be assumed that θ_{n+1} has zero correlation with the other θ_i. (That the extra term is needed follows from the consideration that nothing can be inferred about possible changes in control policy from data collected while conforming precisely to the current version.)

The modelling expression to predict h as $\hat{h}$ would normally be formed as a weighted sum of the θ_i $(0 \le i \le N)$ but because optimization has to depend on differentiation with respect to c, it is necessary to let c appear explicitly. The modelling equation can be written:

$$\begin{aligned}
\hat{h} = &b_{00}\theta_0 + b_{01}\theta_1 + \ldots + b_{0n}\theta_n + \\
&b_{10}\theta_0 c + b_{11}\theta_1 c + \ldots + b_{1m}\theta_n c + \\
&b_{20}\theta_0 c^2 + \ldots
\end{aligned} \tag{4.15}$$

The expression must be of at least second degree in c if there is to be a finite maximum, and one of second degree in c and the θ_i ($1 \leq i \leq n$) will be considered. That is to say, Eq. (4.15) will be terminated at the last term shown, with multiplier b_{20}. The restriction to second degree in these terms does not imply that the model is restricted to second degree in terms of the original signals $a, b, c \ldots$ because the f_i, and the θ_i derived from them, are allowed to include arbitrarily higher-order terms. To let the extremum located be a maximum rather than a minimum, it is necessary that b_{20} be negative.

The optimal control function c is obtained by equating to zero the derivative of Eq. (4.15) with respect to c. This gives:

$$c_{opt} = -(1/2b_{20}) \sum_{i=0}^{n} b_{1i}\theta_i = \sum_{i=0}^{n} k_i'\theta_i \tag{4.16}$$

where

$$k_i' = -(1/2b_{20})b_{1i}. \tag{4.17}$$

The coefficient of $\theta_i\theta_{n+1}$ in the truncated Eq. (4.15), after substitution for c from Eq. (4.14) is

$$b_{1i} + 2b_{20}k_i$$

and hence from Eq. (4.11)

$$b_{1i} + 2b_{20}k_i = \overline{h\theta_i\theta_{n+1}} / \overline{(\theta_i\theta_{n+1})^2}. \tag{4.18}$$

Combining this with Eq. (4.17) gives

$$k_i' = k_i - \frac{\overline{h\theta_i\theta_{n+1}}}{2\overline{(\theta_i\theta_{n+1})^2}b_{20}}.$$

Because $b_{20} < 0$, this can be written:

$$k_i' = k_i + \frac{\overline{h\theta_i\theta_{n+1}}}{2\overline{(\theta_i\theta_{n+1})^2} |b_{20}|}. \tag{4.19}$$

Models

Equation (4.19) shows that when a model is used, the change to be made in a weighting (from its old value of k_i to the optimal value k_i') is proportional to the correlation of h with $\theta_i\theta_{n+1}$. Where f_i was uncorrelated with the other inputs and is therefore equal to θ_i this is equivalent to the use of an "effective fluctuation" computed according to Eq. (4.7), because θ_{n+1}.corresponds to c_f. This shows there is close correspondence between alternative methods devised with and without reference to a model. One difference between the methods is that the model-based method yields a precise magnitude for the changes in the k_i rather than just an indication of their direction.

Orthogonalisation is perhaps less significant than it has been made to appear. If the f_i are treated as mutually uncorrelated and therefore equivalent to the θ_i when in fact they are significantly correlated, the changes in the k_i indicated by Eq. (4.19) are likely to overshoot the optimum, and the continuing optimisation process may hunt without settling and may become wildly unstable. Instability can be avoided by effectively reducing the gain of the optimisation loop. If the adjustments of the k_i are made at intervals, this would mean that the changes made at each step would be a fraction of the indications of Eq. (4.19), or if the optimisation is made to operate continuously, an equivalent reduction in responsiveness is applied.

Difficulties can also arise if the situation differs in other respects from the assumptions implicit in the treatment, as for instance if some of the relationships between h and the k_i are strongly nonlinear, perhaps with discontinuities. Here, too, a reduction in responsiveness may avoid instability. For operation in an unexplored environment, the apparent difference between the model-based and non–model-based approaches, in that the former gives an indication of magnitudes of changes, becomes less apparent.

Reduction of responsiveness, or loop gain, will eventually restore stability, but at the cost of slowing progress towards the optimum. Ideally the controller should operate just a little short of instability, and in Chapter 6 a suggestion is made as to how this might be achieved in a living system.

The suggestion that modes of operation with and without a model are essentially equivalent is difficult to reconcile with intuitive views of how biological systems operate. The behaviour of highly intelligent systems (of which we usually take ourselves as examples) is difficult to describe except in terms of internal models or at least images. A person can sit in an armchair and imagine himself engaged in some activity that can range from climbing a mountain or changing a wheel on his car to asking his bank manager for a loan. The mental images may contain quite a lot of detail, but there is also much that is excluded although it must be evident in the real-world situation. The details that do appear are relevant to some purpose of the imaginer, so even here there is a sense in which the processing is purpose-directed.

The word "model" is used with a variety of different meanings (George 1986), and obviously the internal model formed in the brain does not have the same relationship to what is modelled as a model of, say, a railway locomotive has to a real locomotive. The internal model exists in some physical form quite different from that of its prototype. Models having this abstract character are quite common in other contexts, as we speak of mathematical models, and computer simulations that are also models.

Such models are useful because in some significant way their behaviour corresponds to that of the prototype. They are effectively theories as discussed in Chapter 1, where it was observed that all that an acceptable theory allows us to "know" about the environment is that it is such that observations have conformed to the theory. According to this view, there can be no firm distinction between a control scheme devised with explicit reference to a theory or model and one devised simply by adjustment or "tuning" to achieve a particular response. The "knowledge" embodied in the latter, to the effect that the environment responds in certain ways to certain inputs, can only differ from a model in perhaps being seen as less comprehensive. This cannot be a sharp distinction.

Error Decorrelation

For optimisation with an odd objective function, as in "learning with a teacher" or modelling or prediction, a successful approach has been described as "error decorrelation" (Donaldson 1960). A quantity termed error e is available and can be of either sign, and the requirement is to adjust the multipliers in an expression like that of Eq. (4.2) so as to minimize e. Where e is positively correlated with one of the f_i, it can be reduced by reducing the value of the corresponding k_i, and conversely when the correlation is negative. This is neatly described as "error decorrelation", and Donaldson has applied the principle successfully in yet another approach to the inverted-pendulum balancing task. The learning was, literally, with a teacher, and the system learned to balance the pole by copying the responses of a human performing the task.

Error correlation can be justified in terms of regression. It is very simple if it is assumed that the f_i are uncorrelated (and, in order to be uncorrelated with the constant term f_0, are adjusted to have zero means), though applicability of the rule does not depend on this. The k_i values, now as k_i' to represent "new" values, are evaluated according to Eq. (4.11):

$$k_i' = \overline{f_i y} / \overline{f_i^2}. \tag{4.20}$$

Values of y can be derived from the error e as

$$y = \sum_i k_i f_i - e \tag{4.21}$$

where now the k_i are "old" values. Substituting in Eq. (4.20),

$$k_i' = \left\{ \overline{f_i \sum_j k_j f_j} - \overline{f_i e} \right\} / \overline{f_i^2} = k_i - \overline{f_i e} / \overline{f_i^2}, \tag{4.22}$$

where the removal of the summation sign is possible because the f_i are assumed uncorrelated and only the term in f_i^2 is nonzero. Equation (4.22) is a formal representation of "error decorrelation", derived for the case of uncorrelated inputs.

It is interesting to note that the famous "perceptron training algorithm", described in many works and very clearly in Nilsson (1965), is essentially a matter of error decorrelation. A perceptron is a scheme for pattern classification, in which applied patterns are presented on a layer of sensory units and are then processed by association units to produce a new set of signals that are functions of the inputs. There is then a set of response units, one corresponding to each pattern category to be recognised.

Each response unit is a threshold element giving a binary response of "true" or "false", depending whether a weighted sum of signals from association units reaches some threshold. The signals from the association units are usually assumed to be binary. The purpose of the training algorithm is to adjust the weights so that the unit responds only to the intended pattern category. In the training period a succession of patterns is applied, with, for each, an indication of the correct classification. If a unit responds correctly (gives "true" if the pattern belongs, or "false" if it does not), no change is made in the weights. However, if the response is "true" where the pattern does not belong, the weights associated with those association units that were active are reduced, and conversely if it is "false" for a pattern that belongs, the weights for the active set of association units are increased. If a numerical "error" is assigned value zero for a correct response, and respectively $+1$ and -1 for the two types of error, this can be seen as error decorrelation.

Self-Organisation

An adaptive system of the kind described is particularly interesting if it can automatically add and remove terms from the expression on the right of Eq. (4.1) as well as adjust numerical values. A system allowing this has a claim to be described as self-organising, and a great deal of early literature shows that this characteristic, though rather loosely defined, has been thought to be an essential feature of a nervous system. At least four early conferences had titles that made reference to it (Yovits and Cameron 1960; von Foerster and Zopf 1962; Yovits, Jacobi, and Goldstein 1962; Garvey 1963) and it occurs in a more global or cosmic context in later works (Jantsch 1980; Prigogine and Stengers 1984).

One way, probably among many, in which usage of the term is inconsistent is in the extent to which it is assumed to imply improvement of performance in pursuit of some goal, as represented here by reference to hedony increase. The early discussions were rather closely related to theories about biological functioning, especially in the nervous system, and generally involved an assumption of goal-directedness. Much of later work, particularly the treatment of "dissipative structures" by Prigogine and his followers, has indicated how ordered structures can arise spontaneously but without regard to function. The spontaneous appearance of order can only lead to improved performance if it is combined with a means of selection related to performance. The distinction is discussed by Corning (2003) who suggests that only systems that show improvement should be described as self-organising. He refers to the others as "self-ordering".

An obvious condition for elimination of a term from an expression such as the right of Eq. (4.1) is that the value of its multiplier should remain close to zero for sufficient time, though there is some vagueness over what should be considered "close" and what should be sufficient time. An indication of the former could be given by the magnitude of fluctuations in the value over some period, though this again leaves open the matter of deciding what is a sufficient period. Description of a system as self-organising has to be somewhat intuitive, especially as a given change could be seen by one observer as formation (or dissolution) of a connection but by another as the alteration of a gain setting from (or to) a value close to zero.

Conditions for insertion of new terms should presumably be similar to those for adjusting the values of existing terms. One possibility would be to introduce terms "on approval" by some random process and then to let them be eliminated if they produced no improvement. However, there are ways of choosing potentially useful terms more systematically. For operation with an odd objective function, a new term can advantageously be introduced if its variable is strongly correlated with the error, or with the "correct answer". Variables in the new terms can be either new to the computation or combinations of existing terms.

In the even-objective-function, or hill-climbing, situation, the criterion for introducing a new term is slightly more complicated and must depend on the averaged triple product of hedony, the candidate variable, and the superimposed fluctuations, corresponding to the numerator on the right-hand side of Eq. (4.19).

A vigorous process of self-organisation by either of these means must depend on evaluation of many correlations between variables. This could be achieved by having many correlation detectors in operation simultaneously, or by having a relatively smaller number of "wandering correlators" making spot checks. The latter is clearly more economical, but the former is quite feasible in a nervous system. That something of the sort does in fact operate is shown by the phenomenon of the conditioned reflex, where a signal that had no prior connection with a particular function comes to be incorporated as a result of correlation. It is possible that the new signal is particularly likely to be incorporated if it allows a faster response, as it does in the conditioned-reflex situation, but this is an additional consideration.

The problem of finding indicators for the insertion of potentially useful terms arises also in applying the statistical procedure of regression analysis. This corresponds strictly to the odd-objective-function situation, but there may be a way of adapting the suggested methods to the other. Ways of choosing new terms on the basis of residuals from previous fittings of a regression equation are treated by, among many others, Draper and Smith (1966).

An aspect of optimisation that has not been considered up to now is the means of forming the statistics employed. Their derivation as "running values" is considered in an appendix to this chapter, and the processing of successive batches would also be possible. In either case, a choice must be made of the amount of history to be taken into account, either by choice of batch size or of the time-constant of smoothing, with the latter presumably more relevant to biological adaptation.

For a controller that is self-organising in the sense of being able to introduce new variables to the control function, it is likely that the optimal time constant will increase as the self-organisation continues. For example, a process plant may be affected by the cooling effect of the wind, and if the controller has no information about this (wind direction and strength and air temperature), the necessary compensation has to be provided by the adaptive mechanism, whose response must be fast enough to allow this. However, if the necessary atmospheric data is incorporated in the control equation, the time-constant can advantageously be increased to allow the detection of more elusive influences and more accurate adjustment of parameters.

JANET

A scheme that attracted a good deal of interest at the time was due to Foulkes (1959) with the name JANET. (The name is not an acronym but simply one that Foulkes liked.) The task is to determine the statistical structure of a continuing sequence of characters, which were binary digits in the version implemented, and thereby to predict the next character to emerge. The system forms a set of units that respond if the immediately preceding part of the sequence conforms to a particular pattern and accumulate counts of the number of times each type of digit has followed that pattern. For example, the system might have a unit responding to 1101, and if it has been found that 1101 is usually followed by 1, the system can make this prediction.

The specially interesting feature is that the system has the means of growing its own structure, appropriate to its input. Each unit collects extra data allowing this. For example, the unit responding to 1101 also collects data about the sequences 01101 and 11101, and if there is a significant difference between the predictions allowed by these, the 1101 unit is replaced by two units responding respectively to 01101 and 11101. These in turn collect statistics for sequences that are one digit longer, and either may eventually be replaced by a further two units responding to these.

Simple though it is, the JANET scheme illustrates a means of autonomous organisation to suit an environment. Though Foulkes did not describe it as such, it can be represented as making its predictions by evaluating a polynomial, and the introduction of new terms is triggered by appreciation of correlations in a way corresponding essentially to what has been described. The fact that a unit is eliminated whenever a new pair is introduced is a feature specific to this scheme, related to the fact that the output is required to be a single binary digit.

The scheme does not use anything in the nature of "wandering correlators" because attention is given only to new terms that can be formed by adding an extra digit to the characteristic sequence of an existing unit. This means that there can be statistical structure in the input sequence that fails to be utilised by a JANET, and something in the nature of "wandering correlators" could be used to make random checks on longer sequences. Nevertheless the principle has been employed with success in at least one game-playing program.

Checkers

In his famous program for checkers (draughts), Samuel (1963) allowed for automatic selection of terms in his "scoring polynomial" or "static evaluation function" determining moves. The selection was purely among the set of computed properties of the board position devised by Samuel at the outset based on knowledge of the game, along with binary (or pairwise) combinations of these. One of the computed properties, called "piece advantage", was always included, for the reason given below, and there were 38 other terms that could be included. Of these, 26 were calculated directly from the state of play on the board, and a further 12 were formed as binary combinations. Of the total, 16 were included in the scoring polynomial at any one time and the other 22 were in reserve.

The need to have the piece-advantage term always present results forms the nature of the aim of the adjustments. As has been mentioned, this was not, strictly, to minimise an error, but to minimise the discrepancy between the result of applying the s.e.v. directly and that from applying it at the terminal leaves of a lookahead tree and backing-up by the minimax procedure. It is necessary that the procedure does not fall into the state where all the multipliers are zero, which would give perfect agreement between these two evaluations, but in a way that is useless for the purpose. The forced inclusion of the piece-advantage term avoids this. This term is simply a comparison of the numbers of pieces held by each player, and its relevance to play is obvious. Another advantage of including it is that it directs the play in the right direction, so that the program plays to win rather than lose.

The choice of terms was made to depend on correlation, essentially as suggested here, except that it was used to reject terms from the included set of 16 rather than to choose the new ones. A term that had lowest correlation on a certain number of moves was transferred from the active set of 16 to the reserve, and one from the reserve took its place. Persistently poor-performing terms were rejected entirely.

Because the only terms used are those devised at the outset, there is no place for "wandering correlators" unless to find new terms like the binary combinations that introduce nonlinearity in the scoring polynomial. Four kinds of combination were used, though more could certainly be devised. The number of possible pairings of the 27 primaries is 351, and the number of possible terms using the four kinds of combination is the rather daunting figure of 1404. Samuel chose just 12 of these. It can be seen from Samuel's paper that the binary-combination terms played an important part. He gives a list of the twelve most effective terms, out of the 16 that were in operation at the conclusion of a series of games, and of these 6 are primary terms and 6 combined ones.

Samuel acknowledges that the operation would be more interesting, and presumably a better model of human learning, if the program could generate new terms for itself. As he says, no satisfactory scheme for doing this has yet been devised. This raises a highly significant general issue of criteria of similarity, referred to in one important context as "heuiristic connection" that will be discussed in Chapters 6 and 7.

Pandemonium

Oliver Selfridge (1959) described a scheme to which he gave this title to emphasise that it is most readily visualised as involving a great deal of parallel activity. Like the perceptron, it is described as performing learned pattern classification, but with much more attention to mechanisms allowing autonomous improvement of performance. It is a very early proposal for a device allowing automatic selection of terms in a polynomial function and hence of a form of self-organisation. It is discussed further in the next chapter.

Emergence of Concepts

In discussing the intelligence of animals and humans, it is not usual to pay so much attention to continuous adaptation as has been given here. This is despite the fact that in the early days of cybernetics, a good deal of attention was given to such topics as the responses of a human operator in a control situation, for example by Craik (1947) and North (1952).

There are many indications that the distinction between discrete and continuous processing was not fully appreciated by some early workers. For example, it was suggested that the General Problem Solver, or GPS (Newell, Shaw, and Simon 1959; Newell and Simon 1963), was indeed general and might be bootstrapped to improve its own operation. One reason this is unrealistic is that the GPS is applicable only to what are termed "well-defined problems", or problems for which a purported solution can be tested algorithmically for validity. The GPS is unsuited to the gradual improvement in performance that would model biological learning, and one reason for this can be seen as the failure to deal suitably with continuity. (Strictly speaking, the GPS rationale makes use of continuous measures of subproblem utility and subproblem difficulty, but only in an auxiliary role.)

It has been demonstrated in the foregoing that continuous adaptation is far from being the simple matter that is often assumed and allows changes that merit description as constituting self-organisation. This has been demonstrated in the case where the control action is based entirely on current observations, with no assumption of "hidden" memory in the environment. The situation becomes still more complicated when dynamic responses are considered, or when changes of state are taken into account as in the "reinforcement learning" of Sutton and Barto (1998).

The situation has also been simplified by assuming that the degree of goal achievement is indicated by a single quantity, here termed "hedony". Wiener (1948) uses the term "affective tone" and in formal mathematical contexts the term is "objective function". In many situations, many different criteria will need to be incorporated in a useful hedony measure, with respective weightings and probably with some entering in a "veto" fashion and describable as constraints. For manual or machine computation, it is advantageous to treat constraints as such, rather than as part of the objective function, though in principle they can be incorporated.

Another way in which a single measure of hedony is unrealistic is that goal-seeking behaviour often takes the form of a hierarchy of goals and subgoals. Such a hierarchy is a central feature of the GPS program, and others are apparent in biological systems and in everyday life. For a biological system, survival is usually a main goal, though not always necessarily paramount. For much of human and animal behaviour, the survival of the genes appears to be the main goal, to which survival of the individual could be considered a subgoal, and there are many other ways in which behaviour is motivated by goals other than these and possibly overriding them.

Even if attention is restricted to the goal of survival of the individual, it is easy to see that subgoals come into play. At the physiological level, a myriad of regulatory processes can be described in terms of goals, and at the behavioural level there is again a myriad, including as an arbitrary example that of remembering to look both ways before crossing the road. However, the origins of intelligence were presumably in relatively simple environments, and it is probably not too misleading to assume a single goal whose degree of achievement can be termed "hedony".

There is a connection here with the earlier discussion of time-constants of smoothing involved in adaptation. The most obvious kind of hierarchical structure is one in which the adaptation process at the top level sets goals for those immediately subsidiary to it, which in turn set goals for those below. In that case the time-constants must be longest at the top and decrease down the chain, because the top level can only evaluate a change in its policy once the subsidiary elements have had time to adjust to it.

The description of behaviour in terms of goals is necessarily a theory, or construction formed by an observer. It is however one that fits the observations so well that its use can hardly be avoided. An objection that can be made is that it is a teleological explanation of observed phenomena, as it accounts for current occurrences in terms of a future state. The significance of teleology has been discussed by various workers, and notably by Sommerhoff (1950), who suggests that the ability to act with reference to a future state is a defining characteristic of life.

There is a sense in which a teleological explanation is less fundamental than a causal one in which effects parallel the arrow of time. In the early days of cybernetics, this was a point of conflict between the new way of thinking and older traditions. It was wise to eliminate any trace of teleological explanation in a paper submitted to, for example, the *Journal of Physiology*.

Where a teleological explanation is applied, there is usually presumed to be an underlying causal chain of events that can be cited as an alternative. When a person steps from the path of an approaching vehicle, the explanation he would give for his action would be teleological, along the lines of: "It might have hit me if I had stayed where I was". An alternative non-teleological explanation would refer to the evolutionary and environmental origins of the neural capacity to appreciate the undesirability of impact and to extrapolate motion, and neural and muscular events stemming from these in the person concerned.

There is, however, a sense in which the teleological explanation is much more than just a concise way of referring to the complex causal version. As emphasised by Rosen (1985), there is a sense in which the causal explanation "misses the point" in something the same way as description of a book as ink marks on paper is undoubtedly valid, but fails to explain why perusal of the book may be instructive or pleasurable or judged a waste of time. Another point, of course, is that the details of the neural processes playing a part in the causal story, and especially their evolutionary origins, can only be surmised in general terms. The evolutionary origins are likely to stem from conditions and events in the very remote past, possibly several millions or even billions of years ago.

The discussion of the nature of goals and teleology is something of a digression from the main theme of the place of continuous adaptation in biological learning. It has been argued that continuous processing is fundamental, but at the same time the processing of which we are consciously aware has a different character and involves discrete concepts. That is not to say that concept-based processing has superseded the continuous kind, which is still effective, though mainly "behind the scenes" of conscious thought.

As argued earlier, organisms that experience a range of environments, or of drastically disparate situations within an environment, must become capable of classification. This is likely to be the origin of discrete choices and hence of concepts, though it is undoubtedly true that for these evolutionary origins as for others, details can only be surmised.

The set of concepts held by a person is obviously not static. Fresh concepts are formed and old ones forgotten, in a manner analogous to biological evolution. The chance of survival of a concept is related to how useful it proves to be, and useful concepts give rise to others similar to themselves. The aim of Selfridge (1959) in devising his "pandemonium" scheme was to model such a process in the restricted context of pattern classification. As will be seen in later chapters, a major contribution from this work is to pose a number of questions that have still not been answered but are relevant to the current discussion. Following the argument that a concept is retained if it proves useful, it is possible to consider the use of concept-based processing as itself a concept that has undoubtedly proved its worth. This idea has been indicated by reference (Andrew 1981) to the "concept of a concept".

A "concept" has to be understood in a wider sense than might be suggested by consideration of language and formal reasoning. In this wide sense it is not necessary that the entity using the concept be able to name it in any language allowing communication outside itself. For example, the findings of Lettvin et al. (1959) on the frog retina, and the observations of Darwin (1897) on earthworms pulling leaf-like objects into their burrows, both provide evidence for processing that can only be described satisfactorily in terms of concepts. Darwin observed that a worm chooses the more pointed end of a leaf in order to drag it, irrespective of whether it is the stem end or the distal one. He confirmed his theory by putting down shaped pieces of blotting paper and watching how worms dealt with them.

Lettvin and his colleagues examined responses from the frog retina, as passed to the optic nerve. These are not the primary responses from the rods and cones, but the result of processing these in the layers of the retina. They identified four types of response unit, each of which was very plausibly related to the needs of the frog in its natural environment. The interpretation has been challenged by Gaze and Jacobson (1963) whose more prosaic alternative is in terms of the concentric response fields with "on" and "off" and "on–off" responses, in accordance with well-established previous findings. The pros and cons will certainly not be argued here, except to say that the two interpretations need not be incompatible, and it should probably be expected of a living, and therefore evolving, system that at any given stage of the evolution, some features allow alternative valid modes of description.

Bacterial Chemotaxis

The suggestion that concepts originate as a means of classifying environmental states receives possible support from observations of the swimming behaviour of certain bacteria (Berg 1975; Adler 1976; Alberts *et al.* 2002). They swim towards locations having higher concentrations of nutrients, such as sugars and amino acids, and away from locations with higher concentrations of noxious substances. They manage to do this with only two kinds of motor action, one of them straight-ahead swimming and the other a state of "tumbling" that produces random reorientation.

To allow apparently purposeful progression in three dimensions, the bacterium has to be able to distinguish two states of its environment. One is the state in which the concentration of nutrient is increasing, or of noxious substance decreasing, and in this state the favoured action is to keep on swimming. In the other state, with concentration of nutrient falling or of noxious substance rising, the favoured action is to use the tumbling state to change to a new direction. With these simple rules, the bacterium is able to navigate usefully in three dimensions.

Discrimination of states of the environment can be seen to have emerged at this primitive evolutionary level. Whether or not the use of concepts by multicellular organisms, such as earthworms or ourselves, can be considered to have stemmed from this is impossible to say, but it is interesting to note that Alberts *et al.* consider that certain features are carried over from bacteria to higher forms descended from them. These writers say: "many of the mechanisms involved in chemical signaling between cells in multicellular animals are thought to have derived from mechanisms used by unicellular organisms to respond to chemical changes in their environment".

The discrimination of environmental states by bacteria is part of a remarkable overall mechanism and is related to the fact that the bacterium's means of propulsion has evolved to operate in either of two ways. The organism is propelled by helical flagella that are driven to rotate at speeds in excess of 100 revolutions per second. When driven to rotate in one direction, the flagella come into alignment and the effect is to drive the bacterium forward, in its "swimming" state. When they are driven to rotate the other way, they fan out and produce the "tumbling" action.

A human inventor would be proud to have devised such an economical mechanism for producing the two effects. The means of propulsion and the means of sensing and classifying environments have somehow evolved together, perhaps by a fortuitous combination of mutations. The relevance to the examination of concept-based processing by higher animals is undeniably obscure, though the fact that a form of sensory discrimination emerged in an essentially continuous optimisation task is suggestive.

The ability of bacteria to operate in this way in three dimensions has been quoted as contradicting Ashby's "Law of Requisite Variety" (Ashby 1960), which states that a controller must have as much variety as the system it controls, or that the controller must embody a full model of that system. No description of the behaviour of the bacterium would suggest it has appreciation of three-dimensional space and yet it operates successfully within it.

Where a "model" is understood in the functional sense suggested here, the bacterium does have a model of its environment, namely as being such that a certain pattern of behaviour involving swimming and tumbling results in useful chemotaxis. There are obviously other modes of behaviour allowing more efficient chemotaxis, as exhibited for example by white sharks and piranha fish homing on sources of blood, and these are difficult to describe without assuming some sort of awareness of three-dimensional space. The apparent conflict with Ashby's law can be attributed at least partly to his attention to regulation rather than optimisation or goal-seeking. The behaviour of living organisms is more accurately represented by these latter terms than as regulation.

The search strategy of the bacteria has some correspondence to the method of computational optimisation devised by Rosenbrock (1960) and discussed in many subsequent works including Schwefel (1981). This performs optimisation or hill-climbing in a space of arbitrary dimensionality n (so implying dimensionality $n + 1$ if the hill is also represented). It proceeds in a series of stages, where for each stage a principal direction is assigned in the n-dimensional space.

The procedure differs from the bacterial strategy in allowing exploration along each of a set of n orthogonal directions of which the principal one is a member, and in allowing exploration both ways along these. It also differs in assigning the principal direction for the next stage as a function of the advance made during the last one, instead of randomly. It is, however, similar in terminating the stage and reassigning directions when there is evidence that exploration using the current assignment has ceased to be profitable.

Daisyworld

It has been shown by Lovelock (1983, 1986) and Watson and Lovelock (1983) that a regulatory effect can emerge from a system having none of the expected characteristics of a control system and therefore little correspondence to the treatment here. The effect has been postulated as a probable basis, at least in the initial stages of its functioning, for the "Gaia hypothesis", advanced by Lovelock (1979) as a "new look at life on Earth". The name Gaia is one that was given to the Greek Earth goddess, and the hypothesis challenges the common assumption that environmental conditions on Earth just happen to be compatible with life. It is suggested, instead, that these conditions are best seen as being regulated at suitable levels. Regulation is imposed by the totality of life in the biosphere, effectively forming one giant animal with internal homeostasis.

The environment is much influenced by biological activity. Without it, for instance, there would be very little atmosphere because in thermodynamic equilibrium, the bulk of both the oxygen and the nitrogen would be in nongaseous forms. Lovelock suggests various channels through which biological activity might control such environmental variables as the surface temperature of the earth, composition of the atmosphere, and salinity of the oceans. He is able to show that global variables, particularly the surface temperature, have remained remarkably steady over a period in which the sun's power has presumably increased by 25% (assuming that it is behaving as a typical star).

To show how such control could come into being, Lovelock devised a simple model of a planet on which two varieties of plants (referred to as "daisies") grow. One of these is dark in color ("black" for convenience) and the other light ("white"). Proliferation of black daisies leads to increased absorption of solar radiation and hence warming, whereas proliferation of the white variety has the opposite effect. The bare ground of the planet has an albedo between those of the two plant varieties. A computer simulation was used, and an alternative formulation is due to Andrew (1988b).

It is assumed that the growth rate of each variety, as a function of temperature, is single-peaked. The exact shape is not important, but an important point is that the function, including the optimal temperature for growth, is the same for both varieties. In the computer simulation, it is assumed that both species have a death rate that is again the same for both and independent of temperature.

Another essential but not unreasonable requirement is that the two varieties are sufficiently segregated in their growing areas that the local temperature for black daisies is higher than that for white ones. The simulation due to Watson and Lovelock includes simple and plausible assumptions about the thermal conductivity of the planet.

The simulation has shown that this simple model is effective in keeping the mean temperature of the planet close to the optimum for daisy growth, over a wide range of levels of incident radiation. It is not difficult to explain the result — when the incident radiation is such that the equilibrium temperature for bare ground is below the optimum for growth, black daisies have an advantage because they are locally warmer. They therefore proliferate faster than does the white variety and this causes warming of the planet as a whole. The converse applies when the incident radiation corresponds to a bare-ground temperature above the optimum for growth.

Daisyworld has been termed a "parable" to encourage interpretations and analogies well away from the original model. "Daisies" can be replaced by anything that has diversity and a growth rate dependent on an environmental variable, where it itself influences the variable locally and globally but with greater effect locally. The principle fits well with the Gaia hypothesis and of course does not preclude the later evolution of regulation by other means.

Appendix to Chapter 4: Statistics as Running Values

Running Values

Any system to be described as learning (or self-optimising or self-organising) has
to accept a large inflow of data and to extract the relatively small amount needed
to determine changes within itself. Such condensation must be a feature of human
learning and is explicit in the compilation and processing of statistical data. However, the established methods employed in statistical analysis are normally applied
to a chosen batch of data, which might be the values taken by a set of variables over
some interval of time.

On the other hand, the processing involved in informal human learning does not
have this batched nature, and schemes for artificial systems capable of self-tuning
or self-organisation are most readily visualised as using what may be termed running estimates of statistical values. These are continually updated to take account of
fresh observations, in such a way that recent observations are weighted more heavily than are earlier ones. Because the discussion in this chapter depends heavily on
statistical considerations, it is appropriate to give attention to how running estimates
might be formed. The precise mathematical formulation is of course appropriate to
machine computation but has some correspondence to apparent results of biological
processing.

A running estimate of a mean value has been used for sales forecasting by Muir
(1958). Running values are implicit in much of control theory, especially that related
to the topic of Kalman filtering (Sorensen and Sacks 1971; Meditch 1973). Treatment under that heading emphasises the matter of stability of the entire controller.
Means of computing some standard statistical measures as running values have been
derived by Andrew (1988a) and are likely to be useful when the data is strongly cor-
rupted by irrelevant fluctuations, considered as "noise".

Weighting Patterns

Where a statistic is to be computed as a running value, from a sequence of input
values, the latter must effectively be weighted according to some pattern allowing
recent inputs to carry most weight. The usual operation of smoothing, or quasi-
integration, as employed by Muir, corresponds to an exponential pattern.

The two weighting patterns that are computationally manageable are the exponential and the rectangular. For the latter, a certain number, say n, of most recent
observations are weighted equally and the statistics computed from them without
regard to anything earlier. Because of equal weighting within the sample, standard
statistical theory can be applied for such purposes as deciding a level of significance
of a deviation from an earlier value.

The computation of the usual statistical measures depends on forming a number
of summations. These include sums of the raw observations and of squares and

products. For a rectangular weighting function there is, in principle, a reasonably efficient way of updating these sums, provided the most recent n observations are held in a buffer store. For each new observation, each sum must be incremented appropriately and also decremented according to the value displaced from the buffer by the new entry. This method has some disadvantages, apart from the overhead of providing the buffer store. Unless the various summations are "refreshed" from time to time, any error, say from rounding-off of values, will be perpetuated indefinitely.

The use of exponential weighting is robust in the last-mentioned respect and is more convenient and efficient computationally. It is also intuitively more attractive as a model of biological learning. Both methods require the arbitrary choice of a number n that for the rectangle is the number of observations stored and for the exponential case is a time-constant of smoothing. The arbitrariness is less serious with exponential smoothing because it determines attenuation rather than nullification.

Exponential Smoothing

To obtain the effect of exponential smoothing, each running value is updated to take account of each new observation, according to:

$$\bar{x}_i = \alpha \bar{x}_{i-1} + (1/n)x_i \tag{4.23}$$

where $\bar{x}_i$ is the estimate of the mean after taking account of the i-th observation x_i. The factor α is equal to $(n-1)/n$. The contribution of an input n steps back is diminished by the factor α^n, which tends to e^{-1} as $n \to \infty$.

Estimates of variance and of correlation depend on summations of squares or products of variables where the latter have been "corrected" in the sense of having their mean value subtracted. What might seem to be the obvious way to achieve this would be to compute the means of observations by one smoothing process and then to "correct" each observation by subtracting this mean from it. The corrected values would then be used to form the products to be summed. If running values are computed in this way, there is a smoothing process within a smoothing process, with two different time-constants of smoothing and a consequent risk of unwanted transient effects.

For the standard computation of statistics from batches of data, it is not usual to compute the means first and then to subtract them before proceeding. Instead, use is made of the relation

$$\sum_{i=1}^{N} (x_i - \bar{x})^2 = \sum x_i^2 - (1/N) \left(\sum x_i \right)^2 \tag{4.24}$$

for variance, or for cross-product

$$\sum_{i=1}^{N}(x_i - \bar{x})(y_i - \bar{y}) = \sum (x_i y_i) - (1/N)\sum x_i \sum y_i. \qquad (4.25)$$

These rules can be adapted to the running-value situation. Let the result of a smoothing operation as defined by Eq. (4.23), on a variable x, be represented by $S(x)$. It is obvious from the form of the equation that

$$S(x + y) = S(x) + S(y); \quad S(kx) = kS(x); \quad S(k) = k. \qquad (4.26)$$

Analogously to Eq. (4.25) we have

$$S\{(x - S(x))(y - S(y))\} = S(xy) - S\{xS(y)\} - S\{S(x)y\} + S\{S(x)S(y)\}.$$

Because $S(x)$ and $S(y)$ are final values they can be treated as constants, and the right-hand side of this becomes:

$$S(xy) - S(x)S(y) \qquad (4.27)$$

with the desired feature that all the smoothing operations can have the same time-constant.

Digital Precision

Although the formulas of Eq. (4.24) and Eq. (4.25) are often used in statistical calculations, it has been pointed out by Bryan-Jones (1986) that they are not ideal for automatic computation because the result is obtained as a difference of two values that may be rather close, resulting in loss of precision, and this is also a demerit of operation according to Eq. (4.27) for running values. The preferred methods for computing means and sums of squares and products, from a set of N observations, are as follows.

The N observations must be taken in sequence, and for a particular variable, say x, the mean of observations up to the i-th can be denoted by $\bar{x}_i$, with $\bar{x}_N$ the required overall mean. Clearly, $\bar{x}_1$ can be set equal to x_1, and then for i from 2 to N successive values are evaluated as

$$\bar{x}_i = \bar{x}_{i-1} + (x_i - \bar{x}_{i-1})/i. \qquad (4.28)$$

The corrected sum of squares can similarly be denoted by s_i up to the i-th observation, with s_N the final result. Here, s_1 can be set equal to 0, and then for i from 2 to N successive values are evaluated as

$$s_i = s_{i-1} + i(i-1)(\bar{x}_i - \bar{x}_{i-1})^2. \qquad (4.29)$$

This can be extended to a corrected sum of products p_i where p_1 is set to 0 and then successively

$$p_i = p_{i-1} + i(i-1)(\bar{x}_i - \bar{x}_{i-1})(\bar{y}_i - \bar{y}_{i-1}). \qquad (4.30)$$

The derivation of Eq. (4.30) and thus of Eq. (4.29) follows from expansions of p_i and p_{i-1} according to Eq. (4.25):

$$p_i = \sum_{r=1}^{i} x_r y_r - i\,\bar{x}_i \bar{y}_i \qquad (4.31)$$

$$p_{i-1} = \sum_{r=1}^{i-1} x_r y_r - (i-1)\,\bar{x}_{i-1}\bar{y}_{i-1}. \qquad (4.32)$$

Subtraction gives

$$p_i - p_{i-1} = x_i y_i - i\,\bar{x}_i \bar{y}_i + (i-1)\bar{x}_{i-1}\bar{y}_{i-1}. \qquad (4.33)$$

The first term on the right-hand side of Eq. (4.33) can be replaced by the expression obtained by solving Eq. (4.28) for x_i, multiplied by the corresponding expression for y_i. Rearrangement gives Eq. (4.30).

It can be seen that, for processing a batch of N observations, operation according to Eqs. (4.28)–(4.30) is mathematically equivalent to the use of Eqs. (4.24)–(4.25) but in practice allowing better digital precision. The methods for running values can also be improved in this respect.

Running values of $\bar{x}$ are computed according to Eq. (4.23) which can be rearranged

$$\bar{x}_i - \bar{x}_{i-1} = (1/n)(x_i - \bar{x}_{i-1}), \qquad (4.34)$$

and similarly for $\bar{y}$.

A running estimate of the uncorrected mean of products $S(xy)$ is formed sequentially as

$$S_i(xy) - S_{i-1}(xy) = (1/n)(x_i y_i - S_{i-1}(xy)). \qquad (4.35)$$

From Eq. (4.27) we have, using $\bar{p}$ to represent the mean corrected product:

$$\bar{p}_i = S_i(xy) - \bar{x}_i \bar{y}_i \qquad (4.36)$$

$$\bar{p}_{i-1} = S_{i-1}(xy) - \bar{x}_{i-1}\bar{y}_{i-1}. \qquad (4.37)$$

Eliminating $S_i(xy)$ and $S_{i-1}(xy)$ from the last three equations and again using α to represent $(n-1)/n$ gives a formula for updating of the running estimate of a corrected product:

$$\bar{p}_i = \alpha\, \bar{p}_{i-1} + (n-1)(\bar{x}_i - \bar{x}_{i-1})(\bar{y}_i - \bar{y}_{i-1}). \qquad (4.38)$$

This is preferable to the direct use of Eq. (4.27), just as Eq. (4.30) is preferable to Eq. (4.25).

Summary of Chapter 4

Adaptation in a continuous environment has received much attention under the heading of control engineering as well as in a number studies specifically related to biological learning. Attention is given here mainly to nondynamic control, that is, to the case where the control action is based entirely on current observations, with no assumption of "hidden" memory in the environment, and with an immediate indication of "hedony" or degree of goal achievement.

A distinction is made between two types of continuous environment for optimisation. In one the available feedback is an indication of the degree of goal achievement, with no indication of how the control action might have differed to produce a better result. This is the case of blind hill-climbing. The other type of environment arises when error information is available, as in any case of learning with a teacher, or where the aim is to form a model or to make a prediction.

Effective operation in the blind hill-climbing case can be achieved either by making experimental changes in parameters and computing correlations so as to converge on an optimum or by forming a model and then operating on it to derive the optimal state by equating derivatives to zero. It is shown that the two methods are practically equivalent provided that, in the first option, a particular relationship is assumed between fluctuations superimposed on the control outputs and effective fluctuations ascribed to parameters in the control equation. The significance of the result is, first, to justify use of this relationship, and also to support the general conclusion that there is no firm distinction between a procedure devised or described using a model and one based on direct interaction.

Adaptation by the means discussed is not restricted to parameter adjustment but can include automatic means of altering the set of terms included in the control function. Ways of achieving this are discussed, and reference is made to a number of schemes described elsewhere allowing it. Schemes embodying the feature can be described as self-organising and correspond to the phenomenon of the conditioned reflex.

It is acknowledged that the techniques for continuous adaptation account for only a small, but arguably vital, part of intellect. As argued earlier, discrete concepts can be seen as resulting from the need to distinguish environments. Concepts, once formed, are retained or discarded according to their proven utility, and the principle of using concepts can itself be described as the "concept of a concept".

An early evolutionary development that can be considered to show emergence of a simple concept is described. It is a means of navigation employed by certain bacteria in swimming towards a source of nutriment (or away from one of something noxious). In the nutrient case, the bacterium is able to distinguish the state "concentration increasing" from that of "decreasing" and correspondingly either continue swimming or engage in a "tumbling" action that produces random redirection. Whether this can be seen as a genetic precursor of advanced learning and intellect is doubtful, but it may usefully demonstrate a mechanism.

Mention is also made of a means whereby a regulatory effect can emerge from a system having none of the expected characteristics of a control system. It has been postulated as a probable basis, at least in the initial stages of its functioning, for the "Gaia hypothesis", advanced by Lovelock as a "new look at life on Earth". Regulation stems from different growth rates of varieties of some species. The theory fits well with the Gaia hypothesis and of course does not preclude the later evolution of regulation by other means.

In an appendix, means are developed for computing statistics as "running values" rather than from finite batches of data. The precise mathematical formulation is appropriate only to machine computation but has some correspondence to apparent results of biological processing.

Chapter 5
Backpropagation

Learning in Nets

Following the initiative of McCulloch and Pitts (1943), there has been much speculation about the achievement of artificial intelligence using networks of model neurons. The advent of the "perceptron" principle (Rosenblatt 1961; Nilsson 1965) crystallised something definite and functional out of a mass of diffuse speculation, but it is not difficult to show that the "simple perceptron" has limited capability (Minsky and Papert 1969). This can be attributed to the fact that all of the changes in weights constituting its learning are restricted to a single functional layer. The simple training algorithm is possible because all the places where changes occur are in this one layer and contribute directly to the output of the device, but the range of tasks that can be learned is drastically limited.

The learning schemes discussed in the past chapter have the same limitation, whether they depend, like the perceptron, on an odd objective function, or are hill-climbers depending on the even sort. Much more is needed in order to produce really useful learning machines, and a very great amount more if biological performance is to be explained or mimicked.

In the heyday of enthusiasm for the perceptron idea, there were various claims to have produced systems that would learn from experience such general principles as that images were invariant under translation, dilatation, and rotation. (In much of the early work, attention was restricted, for simplicity, to recognition of two-dimensional patterns, and particularly of printed or hand-formed characters. This is obviously a severely restricted viewpoint on visual perception but it was found to throw up quite enough problems to be going on with. The term "pattern classification" is probably more suitable than "pattern recognition" in this context, but either can be used, depending on whether a "pattern" is understood to be a presented image or one of the recognition targets such as a letter of the alphabet.)

The claims for automatic learning of invariances were not substantiated, and in retrospect it can be seen that it was probably unrealistic to expect that they might be. Such capability would exceed what is done by biological perceptual systems within the lifetime of an individual, though obviously not what has been achieved by evolution. At a much more modest level, however, it may be reasonable to expect that

A.M. Andrew, *A Missing Link in Cybernetics*, IFSR International Series on Systems
Science and Engineering 26, DOI 10.1007/978-0-387-75164-1_5,
© Springer Science+Business Media, LLC 2009

where a system is trained to recognise, say, block capital letters, it should be able to come to use, for example, the fact that a useful feature to respond to is the presence or absence of a horizontal bar across the top of the character. Practical classification schemes utilise features of this sort, devised and built in by their designers as "feature detectors". A neural net that will form its own useful feature detectors in a learning situation must be capable of making adaptive changes not restricted to a single functional level.

Strictly speaking, the perceptron as first described can be said to have feature detectors of a sort, as the sensory units, corresponding to the rods and cones of a retina, were described as having random connections, more numerous than would be needed for one-to-one linking, to the next layer of "association units". An association unit therefore receives a signal that combines elements from several locations in the presented image and so can be considered to represent a feature. It is extremely unlikely that features formed randomly will prove useful in nontrivial tasks.

An extra feature has been incorporated in a perceptron variant due to Roberts (1960), in which the allocation of inputs to the association units is initially random, but for any such unit whose output comes to be little used in the subsequent recognition process, the pattern of input connections is dissolved and a new random one substituted. That the output is little used is indicated by a low or zero value of weight for the synapse or synapses by which it is connected to the response unit or units. This added feature allows the adaptation to be much more flexible and is a simple example of the feedback of "worth" that is central to Selfridge's "pandemonium" scheme discussed later.

The simple perceptron can only learn discriminations that are linearly separable in the multidimensional space whose coordinates are the responses of the association units. It follows that, in terms of these inputs, a perceptron cannot learn the exclusive-or (one or the other but not both) function of two binary inputs, and learning of more complex discriminations is correspondingly limited. The limitation can only be overcome by recourse to a net not restricted to a single functional layer, and therefore having "hidden units", a term used to indicate units that are neither inputs nor outputs. To let these be adjusted appropriately, something more than the standard perceptron training algorithm is required. An effective method using what has been termed a "Boltzmann machine" is described in the next section, and the examination and evaluation of other stratagems is a central topic in the current discussion.

Other schemes for learned pattern classification include that of Uhr and Vossler (1963), who also review much earlier work. Their scheme requires more built-in structure than do either the simple perceptron or Selfridge's "pandemonium", but these authors defend it on neurological grounds, suggesting that the necessary structure corresponds to that of the living retina. They refer to the findings of Lettvin *et al.* (1959) on frog vision. The various studies relating to visual perception all have the characteristic that a set of pattern classes is assumed to have been already established. This, with its implication that learning is necessarily of the "odd objective function" kind, implies that these methods and theories do not account for the ultimate origins of adaptive behaviour.

The difficulty of making useful adjustments throughout a multilayer net, or a net not restricted to a layered structure, is acknowledged by Minsky (1963) as the "credit-assignment problem". In the biological context, Horridge (1968) asked: "How does a synapse know when to change?" It is certainly not obvious how changes in a complex net can be determined so as to improve its performance of a complex task, with the changes driven by feedback of degree of goal achievement or "hedony". There are several ways in which this might be achieved or in some way facilitated.

Multilayer Operation

One way in which feedback of hedony could determine changes in a net would be to let the net undergo random changes in a trial fashion and to retain a change only if it produced, over some trial period, an increase in hedony. This would correspond to a naive view of Darwinian evolution depending on mutations. The obvious difficulty with such a method is that it is unlikely, in any feasible amount of running time, to find any useful state of organisation, such as the already-mentioned one in which it would respond specifically to horizontal bars near the top of images.

Very early proposals for multilayer adjustment were made by Widrow and Smith (1964) and by Stafford (1965). Both of them depend on gradually altering parameters of the network, such as thresholds of its elements, until a required output is produced for a given configuration of inputs. The necessary operation of the network in isolation, for the purpose of adjustment, makes these schemes unattractive either as models of biological learning or for use in artifacts. They can be said to depend on "dummy runs".

Local Goals

A number of other theories that have a bearing on the means whereby a multilayer net may autonomously improve its performance of some task involve the pursuit of goals that are not directly related to the feedback of hedony. For example, the Boltzmann machine (Hinton and Sejnowski 1986; Aleksander and Morton 1995) operates to minimise a function that has some correspondence to energy in a physical system. Its neural units are assumed to be firing asynchronously all the time, in contrast to the disciplined feed-forward of perceptron-type schemes. The state of continuous activity agrees with observations on live brains.

The type of network used to demonstrate the Boltzmann-machine principle has, in common with the nets described by Hopfield (1982), the rather curious requirement that all synaptic connections are symmetrical, in the sense that if w_{pq} is the synaptic weight of a connection from neuron p to neuron q, then for all connected neuron pairs $w_{pq} = w_{qp}$.

The two types of net differ in the response functions assumed for their constituent neurons, because a Hopfield net has neurons with precise threshold or step-function behaviour, whereas the Boltzmann-machine proposal requires a sigmoid characteristic. It in fact uses the technique termed "simulated annealing" whereby the sigmoid characteristic has initially a relatively gentle slope, which becomes steeper as learning proceeds. Changes in synaptic strengths occur more readily when the slope is gentle, and the increase in steepness produces relative stability that corresponds to cooling of an annealed metal.

The rules for synaptic modification, in each kind of net, can be represented as minimisation of overall functions that can be termed "energy", but can also be justified by their conformity to the principle for synaptic change due to Hebb (1949). That the Boltzmann-machine method is effective, for instance in learning the exclusive-or function, can readily be confirmed by computer simulation. Means of doing it in that simple context are outlined by Aleksander and Morton (1995).

Another internal goal that has been postulated for neural processing is the minimisation of redundancy (Barlow 1959; Andrew 1973). The simple observation that biological systems respond strongly to fluctuations in stimulus intensity, with steady values apparently less important, can be interpreted as evidence of this. Fluctuations may be with respect to time or with respect to distance. The strong response to changes in time of stimulus intensity is invariably seen in the patterns of action potentials from sense organs, with an initial burst of activity after the change, and some units responding exclusively to changes. An enhanced response to changes over distance is apparent in the contour-enhancement that is a feature of biological image processing.

Redundancy reduction should allow better utilisation of nervous pathways and their metabolic demands, but certainly cannot by itself determine the changes needed to improve the classification performance of the net. It may be worth observing that something similar arises as recognition of a need for succinctness in the higher-level processing that is the concern of AI (Banerji 1969; Andrew 1978).

Much of biological information processing, as well as its intended simulation in AI programs, is compellingly described in terms of hierarchies of goals and subgoals. Early AI studies under the heading of General Problem Solver (Newell, Shaw, and Simon 1959; Newell and Simon 1963) placed great emphasis on this. These were mainly in the context of well-defined problems, by which is meant problems such that any purported solution can be tested algorithmically for validity. In this important respect they are not immediately applicable to the relatively primitive manifestations of purposive behaviour to which the current treatment applies, though some basic principles carry over. (The distinction does not seem to have been fully realised by the originators of the GPS, who suggested the "bootstrapping" possibility of applying it to itself as a program, to improve its performance.)

Hierarchies of goals and subgoals, or at least phenomena that are most readily described in such terms, are very widely evident. The overall goal of survival of living organisms involves subgoals of obtaining nutrition and of maintaining temperature and other parameters at favourable levels. The pursuit of these in turn involves lower-order goals, such as seeking shade if too hot or exposure to sun if too cold, and so on. However, consideration of goal structures of this sort does not explain the emergence of complex behaviour of the kind considered in the previous section, which is envisaged as occurring at a single level, and probably independently at several levels.

Goals can however be nested in more subtle ways than by letting some be subservient to others. The means of modifying a network or other system, so as to improve its performance as indicated by hedony, can be likened to a critic and referred to as such. There is correspondence here to the notional separation of a learning system into an operative automaton and a learning automaton, as discussed by Glushkov (1966). The critic, or learning automaton, can itself be capable of adaptation, so can be seen as a subgoal. The use of adaptive critics is treated in detail by Barto (1990) and Werbos (1990a, 1990b), with a practical application to the widely studied pole-balancing problem given by Barto, Sutton, and Anderson (1983).

The function of the adaptive critic element, or ACE, in the last-mentioned study can be understood in terms of the approach to "reinforcement learning" presented by Sutton and Barto (1998). In this, the environment is thought of in finite-automaton terms, so that an action applied to it in a particular state produces a (perhaps stochastic) immediate reward and a (perhaps stochastic) change of state. The long-term return from the action is the sum of the immediate return and that achievable from the new state. The return available from each state, or "value" of the state, is therefore evaluated by a form of bootstrapping, as the value of any one state depends on the values of others that can be reached from it.

The value depends, in fact, both on the values that can be reached and on the control policy according to which actions are selected as a function of state. The general idea is the basis of dynamic programming, but, unlike it, the learning procedure operates on an unknown environment. It has to estimate state values (or in some versions values associated with state-action pairs) and from these values to derive favourable policies. Because the state values are functions of policy, and policy improvement depends in turn on them, policies and valuations have to evolve jointly. Algorithms are described by Sutton and Barto that allow this, and for most of them they give assurance of existence of a convergence proof. The means of evaluation in the checker-playing program of Samuel (1963) has a similar bootstrapped character and is acknowledged as a valuable and remarkable early contribution.

The approaches postulating subgoals of various kinds, especially the last mentioned, enormously enhance the power of machine learning schemes, as demonstrated by Sutton and Barto's reference to a program for backgammon that really does learn to play at champion standard. Nevertheless, although they have been shown to be very widely applicable, they do not provide the bridge between continuous and concept-based processing that is believed to be the key to the ultimate achievement of Artificial Intelligence. This will be discussed further when some other stratagems have also been reviewed.

Significance Feedback

The already-mentioned schemes of Widrow and Smith (1964) and Stafford (1965) do not provide satisfactory mechanisms for making useful adaptive changes in a net more complex than a single layer. Another possibility is to suppose that the neural net has what was termed by Andrew (1965, 1973) "feedback of significance" along pathways parallel to those by which the main, or forward-going, information flows through the net, but in the opposite direction. The suggestion is that these have evolved as features of living systems. If the network has variable structure, one function of such feedback could be to preserve useful pathways through it.

That something of the sort must be effective in real nervous systems can be argued on very general grounds. A neuron playing a part in, say, raising a glass of water to the lips, is in a highly regulated environment where it normally has no direct exposure to the thirst that the action is meant to alleviate, and yet its operation is modified with practice to make the action more smooth and efficient. Though the details are far from clear, there is certainly feedback that facilitates such adaptation.

Besides preserving useful pathways, the back-growing network can also allow adaptive changes throughout a net not restricted to a single layer. A single-layer net can be adjusted using a simple "training algorithm" because the sensitivity of the output to changes in the contributing units is apparent. For a more complex net, the contribution of a particular unit may depend on transmission through one or more others, also subject to adaptive alteration. What is needed is "sensitivity analysis" allowing evaluation, at points throughout the forward-going network, of the sensitivity of the output to changes at that point.

The principle is illustrated by Fig. 5.1, which shows an element in the forward-going path that performs multiplication of its two inputs. A feedback sensitivity signal $s(t)$ in a path associated with the output gives rise to sensitivities $s(t)b(t)$ and $s(t)a(t)$ associated with the inputs. The sensitivity measure associated with the output of the net would have value unity, feeding back into the back-growing network.

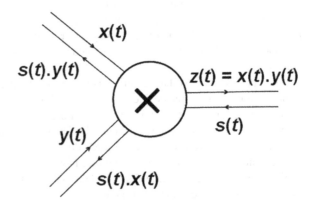

Fig. 5.1 Multiplication element with backpropagation, or significance feedback, of signal $s(t)$

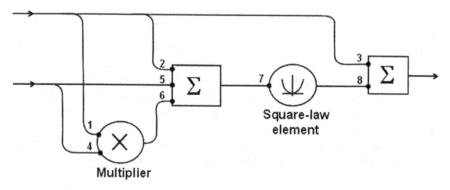

Fig. 5.2 Arbitrary continuous network with backpropagation used in experiments

The signals in the back-growing network must be modified at nodes in ways that are a suitable inverse of the computations in the forward-going paths. For the multiplication element the choice of a suitable inverse is fairly obvious, but less so for other possibilities including the threshold elements postulated as model neurons. The study of continuous networks was undertaken partly as a kind of pilot study that might suggest how to apply similar methods to neural nets, but is important itself in the modelling of learning in general.

To test the method, Andrew (1973) simulated the network of Fig. 5.2, having two inputs and one output. Each point at which a signal pathway enters an element of the net has a variable gain associated with it. These points are shown by numbered black dots in the figure. The output of the adjustable net was compared with that of a similar one having fixed settings for all the gains, and in a succession of trials with random inputs to the network the variable gains were adjusted using significance feedback in a way that was expected to bring the two outputs into agreement.

If the fixed-gain network is termed the "standard", its gain settings and the initial ones for the variable network were as shown in the following table, where the numbers correspond to those adjacent to black dots in Fig. 5.2.

Number	1	2	3	4	5	6	7	8
Standard	0.5	0.6	0.7	0.8	0.9	1.0	1.1	1.2
Initial	1.2	1.1	1.0	0.9	0.8	0.7	0.6	0.5

During a run, a succession of pairs of input values to the networks was produced by a pseudo-random generator, each value being a selection from a rectangular distribution over the range −1 to +1. The "error", or deviation of the output of the adjustable net from that of the standard, was evaluated each time and used in the feedback. Its magnitude was also smoothed with a time-constant of 10 trials and taken as an indication of progress. Starting from value zero, this indication rose to a

value just over 0.2 near the beginning of each run and was then followed until it fell
below 0.02 (or until instability was evident).

After each trial, the principle of significance feedback was applied to obtain, for
all the numbered points, a measure of the sensitivity of the output to activity at
that point. For the multiplication element the method was as described, and for the
square-law unit an obvious modification. For summation units, the feedback signals were the same leaving as entering, except for their modification as they passed
variable-gain points. At one of these points, with current gain value g, both the
forward-going signal and the feedback signal were multiplied by g. The sensitivity
of the final pathway was unity, and other values were calculated back from it.

After each trial of the network, let s be the sensitivity computed in this way for
the output side of a variable-gain point ("output" with respect to the primary signals,
not the feedback), and y is the signal amplitude on its input side, then for each such
point the gain is changed to a new value g' according to:

$$g' = g - k.y.s.e$$

where e is the error, or difference between the output of the variable network and
that of the standard one, and k is a constant.

Because the adjustments at the respective points summate in their effects, k has
to be less than unity to ensure stability, and it is not easy to predict how large a value
can safely be used. A possible mechanism for its adjustment in autonomous systems
will be discussed in the next chapter, with the suggestion that it is likely to be the
method favoured by biological evolution.

Operating as described, with initial conditions as in the table, the network converged, so that the smoothed magnitude of error fell below 0.02, in 344 trials with
$k = 0.05$. Increasing the value of k gave more rapid convergence until instability
occurred for values greater than about 0.086. The critical value was not fixed, as in
one run a value of 0.089 gave convergence to the criterion in 171 trials, whereas in
another a value of 0.086 caused divergence. Consistent behaviour could have been
obtained by arranging that the "seed" of the pseudo-random number generator was
the same for all runs, but such consistency would be misleading.

The experiment as described is of course with reference to error, and therefore to
the odd objective-function situation. The extension to the even objective-function,
or hill-climbing, situation, should be possible by imposing fluctuations on the output and adapting the methods discussed in the past chapter. In the attempt to be as
general as possible, the method was described with reference to a feedback of sensitivity s, but the product $s.e$ in the rule for gain modification corresponds to the
feedback of error that is central to the "backpropagation of error" treated notably by
Rumelhart et al. (1986).

Although the use of backpropagation to evaluate sensitivity, with the values then
multiplied by the overall error, is mathematically equivalent to what is understood
by backpropagation of error, there could be an important difference in practice,
especially in the means that might be imagined for operation of the principle in
biological systems. The overall error (or possibly another multiplier in the even

objective-function case, discussed below) can be broadcast over the net. The evaluation of "sensitivity" is the part of the process that requires such backpropagation through the net elements or through auxiliary structures paralleling them. This separation of the necessarily propagated component from the part that can be broadcast may be significant in connection with some plausible pathways for backpropagation in living systems. As discussed below, these may react relatively slowly, and the delay or lag may be more acceptable when affecting only the "sensitivity" component, than otherwise. This point merits further examination.

Another aspect that deserves further examination is the possible extension of backpropagation to the even objective-function, or hill climbing situation. Adaptation is only possible in this situation if some perturbation is imposed on the net output. According to Eq. (4.7) in the past chapter, the "effective fluctuation" of a multiplying constant (there k_i) is the product of the superimposed fluctuation (there c_f) and the transmitted signal (there f_i). The change to be made in k_i depends on the correlation of this with the hedony h. This is for an element contributing directly to a final summation and therefore having sensitivity one. For other elements in a net not restricted to a single functional layer, the amount of the modification could be modified according to the computed "sensitivity", probably by incorporating it as another factor. This possibility has not been seriously examined by me, and not elsewhere as far as I know. As in the case of operation with an odd objective function, certain components (c_f and h) of the feedback can be broadcast, with only the evaluation of sensitivity requiring backpropagation through net elements.

In the treatment by Rumelhart et al., and in other significant studies of artificial neural nets including that of Hopfield (1982) and the already-mentioned Boltzmann machine, the network elements were artificial neurons differing from the original McCulloch–Pitts version (but not from real neurons) in being able to process continuous signals. The elements sum their inputs from a set of synaptic connections, and then subject the sum to a transformation represented by a sigmoid graph. The transformation is nonlinear but differentiable and so allows the operation of significance feedback or backpropagation. The modelling of neurons as continuous rather than as threshold elements is consistent with observations in experimental physiology, where nerve impulses are usually repetitive in a way that suggests representation of continuous values using frequency modulation.

Rumelhart et al. put the backpropagation algorithm on a firm mathematical basis. They showed that the reduction of error can be represented and analysed as gradient descent. The enormous resurgence of interest in artificial neural nets in recent decades, as a practical computing technique, has largely stemmed from successful application of this algorithm. It has become sufficiently commonplace to give rise to a semi-affectionate abbreviation as "backprop".

Barto (1990) attributes the backpropagation principle to independent developments by le Cun (1985, 1988), Parker (1985), Rumelhart, Hinton, and Williams (1986), and Werbos (1974), to which Andrew (1965) should be added. A closely related idea is implicit in the "pandemonium" principle developed by Selfridge (1959). The prominence given to the precise mathematical formulation by Rumelhart et al. is unfortunate in at least two respects. One is that it focuses attention on optimisation with a fixed structure, rather than on operation to determine structural changes, and the other is that it restricts attention to the "with a teacher" or odd objective-function situation, whose inadequacy has been noted.

In reports of applications of artificial neural networks with backpropagation, it is usual to describe the form of the network that proved sufficient, with perhaps one, two, or more layers of hidden units, and a stated number of units in each layer. There may also be a particular pattern of interconnection, denoted by a term such as radial-basis function. For truly flexible learning, however, the network should include the means of forming an appropriate structure autonomously, either by adding new units and connections to an originally simple form or by starting with a large richly connected net and whittling it down.

In this the Rumelhart method contrasts strongly with Selfridge's proposal, which was aimed at improvement by structural change. Selfridge can in fact be credited with a suggestion for a form of backpropagation that must be the earliest of all, as the feedback of "worth" from cognitive to computational demons.

As with other schemes for learned pattern classification, a "pandemonium" is trained by presentation of a series of sample inputs, along with correct classifications. The means of doing this can be the perceptron training algorithm in any of its variations (though Selfridge describes a hill-climbing procedure). After a period of training, each cognitive demon computes an indication of the "worth" to it of its input from each of the computational demons. In the device as described by Selfridge, the "worth" was simply the magnitude of the synaptic weight of the connection, after appropriate normalisation. The feedback of "worth" from cognitive to computational demons is an important kind of backpropagation.

In its simple form as described so far, the feedback operates between two layers of elements, without any need for the inverse processing of feedback signals that is characteristic of other forms of backpropagation. However, the principle can be extended to nets of computational demons not restricted to a single layer, and inverse processing is then needed. Selfridge's proposal for forming new units by "conjugation" implies the formation of nets not restricted to a single layer, and therefore, though he does not explicitly note it, to inverse processing of the feedback of "worth".

One of the methods proposed for forming new units was "conjugation". This meant taking two computational demons that had proved to have high "worth" and combining their outputs as a new signal to be treated as another computational demon. The function used for combination was chosen randomly from a set of possibilities. This implies that the feedback of "worth" has to operate over a network not restricted to a single layer of computational demons. The two demons contributing to the combining demon will continue to receive their own feedbacks of "worth" as before, but in addition to this they should share in the feedback to the combining demon. The combining demon therefore has to process a back-going signal in the way that is characteristic of backpropagation in other contexts. Selfridge does not comment on this aspect in his paper, but he shows the configuration of demons of Fig. 5.3, which could only be effective when embodying backpropagation that includes inverse processing.

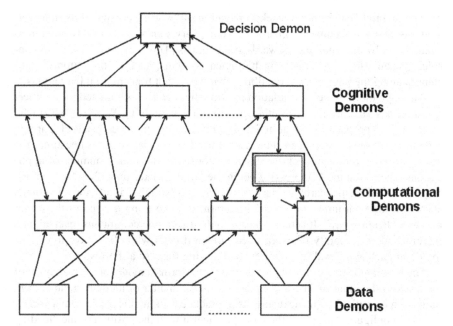

Fig. 5.3 Selfridge's pandemonium with computational demon inserted by "conjugation" and receiving inputs from other computational demons, indicated by double border. Double arrows indicate links that convey both a forward signal and a feedback of worth

The other suggestion for forming new units is "mutated fission" whereby a new unit is to be formed that is similar, in some suitable way, to one that has been highly rated. Implementing this in an artifact poses serious problems because of the difficulty of choosing a suitable criterion of similarity. The aim is of course to simulate biological evolution, which is remarkable in apparently using suitable criteria of similarity, presumably as a result of "adaptation to adapt". Two application of "pandemonium" methods to practical pattern classification are described by Selfridge and Neisser (1963) but without any provision for mutated fission. One application was to the automatic decoding of manually keyed Morse-code transmissions and the other to the recognition of hand-blocked (i.e., drawn by hand as block capitals) characters.

(In fact, neither mutated fission not conjugation was actually implemented in either of the applications described by Selfridge and Neisser, but clear rules for applying conjugation are in the original paper. This work was done when computers were much slower and rarer than now, and the trial introduction of new terms was likely to be done manually between runs rather than automatically. Samuel (1963) similarly introduced new terms in the "scoring polynomial" of his checker-playing program, as combinations of pairs of original terms, but did not provide for their automatic generation.)

An attractive feature of the backpropagation of "worth", compared to other versions, is that the feedback signals need only vary much more slowly (require a smaller bandwidth) than the forward-going signals. This makes it possible to consider transmission mechanisms in biological systems, such as migration of substances along the interior of nerve fibres, that would not transmit rapid variations.

Although the "classic" formulation by Rumelhart *et al.* has attracted most notice, attention has also been paid to other possibilities, notably by Werbos (1989) who refers to backpropagation of utility, and comments on the inadequacy of learning schemes that only function in the "with a teacher" mode. At the same time, the power of the classic approach has been demonstrated in a large number of applications, one of the most impressive being the "truck backer-upper" of Nguyen and Widrow (1990). This learns the task of reversing a (simulated) articulated vehicle, starting from an arbitrary initial configuration, so as to bring the end of the trailer part to a "loading bay". It actually learns to do this in one continuous reversal, and therefore more elegantly than practised human drivers who often have to correct their approach at intervals by stopping and driving forward a short way.

The program learns in two stages. In the first it forms a model of the responses of the truck to the input of steering. Because this is a matter of modelling, an "error" indication is available. The next stage is to practice backing the truck from a variety of initial configurations, using the model to steer it with the position of the "loading bay" as goal. The model is not sufficient to let this be achieved in one go, and a succession of shorter runs is made, ending when the rear of the trailer reaches the line of the loading bay, when an appropriate error is derived as the misalignment with the bay. This error is used by the "classic" method to adjust a virtual net consisting of a concatenation of copies of the modelling net, the respective copies corresponding to stages in the approach. The changes were then condensed onto just one copy of the modelling net, and despite this simplification the method converged on a strategy that, as mentioned, outperformed human drivers.

The "backer-upper" is clearly what is generally understood as a "control" application, processing continuous data. Backpropagation methods have also found favour with linguists, with a variety of applications in their area reviewed and compared by Lawrence, Giles, and Fong (2000). Such studies are essentially concept-based, so that applications of the backpropagation principle fall on both sides of the link indicated as missing in the title of this book. However, the link must still be considered to be missing because, at least up to now, specific applications fall on one side or the other. Backpropagation will be discussed from another viewpoint later, but first its relationship to biological systems will be examined.

Evidence from Biology

There seems to be no direct evidence that significance feedback plays a part in the self-organisation of the nervous system, though the general idea that there should be feedback preserving pathways that are by some criterion useful is very much in keeping with other observed responses of living matter. For example, bones and muscles become stronger with exercise, and "learning to learn" is an accepted psychological phenomenon. For the central nervous system, such feedback could be seen as an evolved partial solution to the "problem of credit-assignment" referred to by Minsky (1963).

It was observed by Singer (1968) that cutting the nerve fibre connecting with a sense organ such as a taste bud causes the organ to atrophy in much the same way as a muscle does when its motor nerve is severed. This appears to support the suggestion that parts of the nervous system are sensitive to the ultimate fate of information passing through them, in accordance with the principle of significance feedback. Unfortunately for such a theory, however, other results reported by Singer, on the effects of innervation on the regrowth of severed limbs of amphibians, suggest that the mere presence of nerve fibres, irrespective of their function, supplies a chemical stimulus to growth. It seems rather likely that the nonatrophying of sense organs and muscles is due to some such stimulus independent of neural activity.

Other apparent, but again probably slightly misleading, evidence comes from work on attention mechanisms by Hernandez-Peon et al. (1956). They showed that the activity in a sensory (auditory) pathway in a cat was greatly enhanced, at the cochlear nucleus, a processing stage well removed from seats of "higher" functions, when the animal was paying attention to the stimulus entering by this pathway. The auditory stimulus (clicks from a loudspeaker) was applied continuously, and the response was reduced when the animal's attention was distracted by seeing live mice behind glass, or smelling fish, or receiving a shock to a paw. This is evidence for feedback to the lower level indicating that information conveyed is receiving attention at the higher level.

As discussed by Horridge (1968), however, these results are not now regarded as conclusive evidence for neural feedback to the cochlear nucleus. The cats were unrestrained and the change in response could have been due to a small change in head position or direction of the animal's ears. (It is very easy to verify that a cat's ears are highly directional and steerable by snapping fingers in various locations around a snoozing domestic cat.) There was indeed a feedback of significance in the Hernandez-Peon experiments, but it probably involved muscular responses, so did not indicate the purely neural effect these workers supposed.

Warren McCulloch has referred in a number of places (McCulloch 1959; Kilmer, McCulloch, and Blum, 1968, 1969) to a principle of "redundancy of potential command" as a feature of neural activity, and particularly of the reticular formation of the brain stem. He suggested that parts of the nervous system behave like a group of men hunting a bear, where the one who can see the bear shouts loudest or most urgently and becomes for a time the leader of the group. He also suggested that the success of Drake's fleet in fighting the Spanish Armada was partly due to a similar policy of letting authority be assumed by the ship in the best position to accept it.

The obvious requirement for this "redundancy" to be useful is an auxiliary feedforward of information, corresponding to the loudness or urgency of the shouts of the respective bear-hunters. In practice, unless the hunters were remarkably well disciplined and similar in temperament (as the captains in Drake's navy may have been), estimates would be formed of the reliability of the respective sources of signals, and so a form of significance feedback would come into play, and it is pretty certain that the same would apply in a nervous system.

A line of investigation that has a bearing on possible neural mechanisms was initiated by Wall and Egger (1971). In their initial study they worked on rats and recorded from an area in the lower brain on which the body surface is mapped. (The term "homunculus" is used for such mappings.) The area is only a few millimetres across, but with microelectrodes they were able to trace out, on each side, mappings of the fore and hind limbs on the opposite side of the animal. For a large number of points in the mapping area, the tip of a microelectrode was placed at the point, and then the rat's body was brushed with medium pressure to locate the spot giving the largest response.

In a number of animals, these workers interrupted, on one side, the pathway connecting the hind limb to the mapping area. If the map was examined within a day or two of the operation, there was a "silent area" where the hind limb would have been represented. When a longer time was allowed to elapse, it was found that after one to three weeks the representation of the forelimb had expanded, as though drawn on a rubber sheet and stretched, into the area where the hind limb had previously been mapped.

There are three main ways in which the effect might be accounted for. One is the possibility that neurons moved bodily, as would happen if neurons in the disconnected area shrank in size and those in the other area swelled. Wall and Egger give reasons for believing this to be unlikely. Another possibility is that new neural connections grew into the disconnected area, and another is that connections that were already in existence, but quiescent, became effective. It is argued, both by these authors and by Chung (1976), that the last of these possibilities is the most likely. Wall also found other situations in which the interruption of a sensory pathway caused the neuron with which it connected to become sensitive to other inputs, often from bodily locations quite distant from the one initially responsive, and sometimes even of a different sensory modality.

However, what seems the obvious explanation of these findings, namely the supposition that a neuron demands a certain amount of input and lowers its threshold if this is not forthcoming, does not fit all the facts. Wall and Egger found that the appearance of the "rubber sheet" effect was dependent on exactly where, and just how, the sensory pathway from the hind limb was interrupted. It did not appear when the interruption was further back than in the initial experiments, and it was dependent on whether the blockage was by severing the nerve or crushing or freezing it. The findings are in fact consistent with the view, advanced by Wall, that the rearrangement producing the "rubber sheet" effect is determined by very fine fibres, termed C fibres, rather than by activity in the main sensory paths. The procedures that produced the effect were those that would damage the C fibres accompanying the main paths. Crushing, for instance, damaged the main fibres but did not interrupt the C fibres because of their small size.

Patrick Wall came to think (Wall, Fitzgerald, and Woolf 1982) that these fine fibres probably play a part in controlling the interconnections of the larger fibre running alongside them. The larger fibres are those whose activity is amenable to electrical recording, and can be related to apparent motor and sensory functions of nerves. No function had previously been ascribed to the C fibres. (The work reported made use of capsaicin, the component that gives chili peppers their fiery taste, which can inhibit the growth of C fibres.)

It is entirely possible that the C fibres could implement some form of significance feedback or backpropagation, but this can only be conjecture. The situation is clearly not simple.

Another possible means by which backpropagation might operate is through microtubules that form a cytoskeleton within neurons and other cells. It has been suggested by a number of workers that these are information pathways, but whether they are fast or slow is uncertain. Jibu and Yasue (1994) suggest that they have the nature of waveguides and offer rapid transmission, which could mean that the central nervous system is functionally enormously more complex than conventional neuron theory suggests. If true, this would make it unsurprising that, as pointed out by Sejnowski (1986), parts of the nervous system perform complex computations with remarkably small computational depth as reckoned by standard neuron theory. A suggestion that the microtubules provide error backpropagation is vigorously defended by Dayhoff et al. (1993) and in a rather different form is also supported by Werbos (1992).

For backpropagation of error, corresponding to what has been termed "vector significance feedback" the feedback signals must have a bandwidth, or capability of rapid variation, similar to that of the forward-going signals. For what has been termed "scalar significance feedback", typified by the feedback of worth in Selfridge's "pandemonium", the feedback signals vary more slowly than the forward-going ones, and other possible mechanisms in biological systems can be considered. The nerve fibre is a tube along which substances can migrate, and slow signals can be conveyed, in either direction, by such migration. Melzack and Wall (1988) discuss the possibility of such communication and refer to speeds of 1 to 240 mm per day for migration of the chemical substances. This is of course extremely slow compared to the propagated nerve impulse travelling at 1 to 100 metres per second.

It should be mentioned that Hinton and McClelland (1988) have argued that, although the backpropagation algorithm has been the basis of the majority of the many practical applications of artificial neural nets, there are reasons for considering it unlikely to operate in living systems. They describe an alternative that they believe to be more relevant as a biological model. They argue in terms of the standard algorithm as described by Rumelhart, Hinton, and Williams (1986). This is described exclusively in terms of backpropagation of error, and hence with reference to the "with a teacher" (or model to be copied) situation. One of the criticisms of Hinton and McClelland refers to this very dependence on the specification of output vectors by a teacher, which severely limits the applicability in real-life situations.

They make a second criticism of backpropagation as being neurally implausible (and also hard to implement in hardware) because of the need for all connections to operate both ways, and for units to apply different input–output functions in the

forward and backward passes. It is difficult to imagine such operation being specified genetically.

These authors describe an alternative form of operation termed "recirculation"
that could come into operation with less demand on genetic specification and is
shown to converge nicely in a particular type of application. Its goal is to generate
an output that matches the input to the net. Such a goal is of interest if the number
of hidden units, through which the information must pass, is restricted to fewer than
that of input–output units. The operation of the net, if successful, then achieves
message compression and hence the detection of statistical regularity in the input.

The application is undoubtedly rather special, and because the required network
output is specified it has to count as "learning with a teacher". In common with
Hopfield nets, and those implementing the Boltzmann machine, the recirculation
method requires synaptic connections to be symmetrical, with $w_{pq} = w_{qp}$.

The observation that the "classic" backpropagation method is neurally and genetically implausible probably has to be conceded, though its force might be lessened if
a modified version could be shown to work with reduced precision, and perhaps also
by separating the feedback determining sensitivity from that of global feedbacks,
as discussed earlier. Also as has been seen, the general principle of backpropagation or significance feedback can take other forms altogether, including Selfridge's
feedback of "worth" which could be effective using the slow channels provided by
migration along fibres, or by transmission along microtubules, according to certain
views of their function.

Although the emphasis on derivatives in Werbos (1994) appears to imply attention to something akin to the "classic" version of Rumelhart et al. (1986), a rather
different impression is given by his claim that the first mention of something to
be termed "backpropagation" was by Sigmund Freud. He also sees support for it
in Eastern religions. These ideas are discussed further in Werbos (1992) in connection with the possible involvement of the microtubules of the cytoskeleton. He
suggests that these might propagate a quantity termed by Freud "psychic energy"
or "cathexis". This is of course highly speculative but seems to imply a spread that
could be relatively slow and similar to the feedback of "worth" as postulated by
Selfridge. The relatively slow transmission of something of this sort is much more
readily reconciled with biology than is the "classic" backpropagation. Possible vehi-
cles include microtubules, diffusion along nerve fibres, transmission by C fibres,
and others that might involve non-neural tissue. (Some theories of memory have
involved the glial cells that are numerous in close association with neurons.)

Structured Feedback

Thus far, backpropagation has been assumed to be of simple numerical magnitudes,
but there can also be feedback having structure. This is easily seen in human interactions, where for example a manager of a business might tell one of his sales-
people that his performance could have been improved by being more forceful, or

more gently persuasive, or more imaginative or sympathetic, and so forth. All of these terms have continuous character, as implied by reference to "more" or "less" of them, even if observers might not always agree about comparisons. It is rather likely that structured feedback also plays a part in internal thought processes, even at the level of hand-blocked character classification, because although the simpler type of backpropagation can be made to determine structural changes and hence to produce useful configurations of hidden units, it is difficult to imagine the emergence in this way of, for example, units sensitive to a horizontal bar at the top of an input pattern, or to totally enclosed white space, though these would contribute to the classification.

The continuous variables effective in structured feedback might be descried as "derived" or "emergent" and have the character epitomised by the "heuristic connection" of Minsky (1963), mentioned at the beginning of Chapter 3 along with other examples such as distance from a solution in problem-solving. Measures of similarity implicit in the use of analogy are discussed by MacKay (1959), Carbonell (1984), and Chouraqui (1985). Carbonell reinforces the point by observing that it is rare for a person to encounter a problem that bears no relation to others encountered or observed in past experience, so that it is at least as important to consider how the thinking process slides from problem to problem as it is to decide how a problem of a totally new kind would be tackled. Minsky's "heuristic connection" between problems amounts to an estimate of the ease of sliding from one to the other as visualised by Carbonell.

The very remarkable capacity of the brain for associative recall also implies use of similarity criteria that have this intangible character. The part they play in evolution and learning was discussed in Chapter 3, where the "basic learning heuristic" of Minsky and Selfridge (1961) and the similar principle advanced by Lenat (1984) are certainly intended by all of these workers to apply to complex situations in which similarity criteria are far from obvious. The difficulty of implementing useful "mutated fission" in a pandemonium arises from that of finding a suitable criterion of similarity for policies embodied in computational demons. In the original paper, Selfridge (1959) accepts that "this is likely to require that the demon be "reduced to some canonical form so that the random changes we make do not have the effect of rewriting the program so that it will not run or will enter a closed loop". However, as he undoubtedly realised, the choice of an appropriate canonical form must in itself constrain the possible criteria of similarity.

The late Anatol Rapoport referred (Rapoport 1982) to "strange topologies" effective in symbolic thought, where a topology implies a metric and hence continuity. (The latter is in the sense that allows interpolation and extrapolation, though in this context not in the sense demanded by differential calculus.) The word "strange" is an acknowledgement of mathematical intractability and it is likely that such topologies can only feasibly be represented in some memory-intensive medium such as a neural net.

"Strange" topologies are contrasted with simple ones that could, for example, refer to similarity of physical representations of signals. The two speech signals "Send reinforcements we're going to advance" and "Send three and fourpence we're going to a dance" are similar if the sounds of the spoken messages are compared using relatively simple criteria but distant in the "strange" topology associated with meaning. (The example is attributed to a party game in which a message is whispered repeatedly from person to person and may be distorted. "Three and fourpence" was understood to mean "three shillings and four pence" in British pre-decimal currency, so the example, though a nice illustration, is seriously out-of-date.) The simplest topologies are those that are functions of continuous physical variables such as the angle of tilt in bicycle riding.

The attribution of associative recall to closeness in a "strange" similarity space is rather contrary to other proposed mechanisms, such as analogies with holography (Gabor 1968; Pribram 1971; Andrew 1997) or the type of neural net due to Hopfield (1982). These, at least in their simpler forms, suffer from being rigid in the type of similarity assumed, whereas the brain can use alternative similarity criteria and can learn or evolve new ones.

(It should be mentioned that Pribram presents experimental evidence that holography, or something like it, is involved in visual perception. The sensitivity of visual systems to spatial frequencies is difficult to explain otherwise and seems to refute the feature-based recognition schemes favoured in AI. However, Pribram makes it clear that he visualises something a good deal more complex than one-stage holographic matching, which might allow the necessary flexibility in effective similarity criteria. It should also be mentioned that, although Pribram's arguments are compelling based on certain evidence, there are also visual phenomena that do not fit.)

The central aim of this book can be rephrased in terms of the "strange topology" idea. It is to examine, necessarily with a good deal of speculation, the process of evolution from simple to strange topologies in living systems, in the hope of imitating it in artifacts. The emergence of classification, suggested in Chapter 2 as the origin of concept-based processing, would be a step in the direction of "strangeness". A further step might be the realisation, in effect, that exclusive classification is not ideal, and that similarities between classes can be exploited. Exhaustive classification in higher animals could be interpreted to mean that the neuron mentioned earlier as being involved in raising a glass of water to the lips would adapt to this action in a way unconnected with adaptations to raising the arm for other reasons, like scratching the head or signalling a bid in an auction. Clearly this does not happen, and it is rather likely that, from the start, there would be ways in which classification is not exclusive.

Nevertheless the realisation, in effect, that distinct entities can have features in common has to be the basis of anything in the nature of reasoning by analogy, whose introduction would be a major step toward "strangeness". There could even be a bearing here on the current "hot topic" of consciousness. A clear evolutionary advantage consciousness offers is the possibility of an organism using itself as a model of others with which it interacts. It could for instance be useful to a predator (or a potential prey) to be able to refer to its own imagined reactions to predict the likely next move of the potential prey (or predator) with which it is involved. This would presumably be most useful to creatures that are sometimes predators and sometimes prey, and probably still more valuable in intra-species conflicts.

This, of course, is extremely speculative, and it is difficult to see how the suggestions might be proved or disproved, but it might be useful as a general viewpoint. A less controversial observation is that processes describable in terms of simple topologies coexist in the brain with concept-based or "higher" or "strange topology" processes, and in Chapter 7 it will be argued that these mirror each other and intelligence can be regarded as fractal. First, in Chapter 6, a number of reasons will be reviewed for believing brain process to be in some sense multilayer or hierarchical.

Another attempt to treat the emergence of intelligence from first principles is the "Laws of Form" of G. Spencer Brown (1969) mentioned in Chapter 2. The making of a distinction, for instance between inside and outside a boundary, has been suggested as the most fundamental component of intelligence, and presumably the main feature of its evolutionary beginnings. Despite the attractive precision of this approach, I feel that this immediate entry to discrete processing does not offer the most fundamental starting point. The precision is attractive to human intellects with a great deal of evolutionary history behind them (though argued here to rely on continuous processing much more than is usually supposed.)

Summary of Chapter 5

Following the lead of McCulloch and Pitts, there were many attempts to make intelligent artifacts as artificial neural networks. A version that aroused interest was Rosenblatt's "perceptron", but its restriction of learning adjustments to a single functional layer meant that its capability was severely limited. The intention was that a perceptron should start with no information about a specific task, and so the "association units" were considered to have randomly chosen sets of inputs from sensory units. In a specific task, such as recognition of hand-blocked characters, the use, instead, of specially devised feature detectors may greatly improve performance. An important part of the task has then been built in rather than learned.

More powerful learning behaviour can be expected from nets whose adjustments are not restricted to a single layer. The difficulty of making adjustments in such a net has been indicated by Minsky as the problem of credit assignment. Early schemes that were proposed depended on "dummy runs" of the network and were both clumsy and biologically implausible. A number of methods that allow changes in pursuit of local goals are reviewed, including the Boltzmann-machine principle that allows learning of discriminations not linearly separable. Nested goals may also be useful, especially the nesting of an "adaptive critic".

A class of methods for overcoming the problem of credit assignment is indicated by "backpropagation" and was proposed by the author in an early note as "significance feedback". In one form it was implicit in Selfridge's earlier "pandemonium". The form in which it is best known and has been widely applied is the "backpropagation of error", which necessarily refers to the odd objective-function, or "learning with a teacher" situation. The extension of the method to the even objective-function or hill-climbing case has not received much attention. A different approach is the feedback of "worth" suggested by Selfridge.

Although the "backpropagation of error" algorithm has been the basis of many practical applications, it is implausible as a feature of biological processing. The feedback of "worth", or of "utility" as suggested by Werbos, is however biologically plausible.

The principle of backpropagation is then extended to refer not only to simple numerical magnitudes but also to feedback having structure, as in human interactions where someone might be told he should have been more forceful, or more gently persuasive, or more imaginative or sympathetic, and so forth. This depends on continuous measures of the type of Minsky's heuristic connection and the "strange topologies" of Rapoport. These ideas are expanded in the following two chapters, first by showing that intellectual processes are best viewed as multilevel or hierarchical and then that they have fractal character.

Chapter 6
Self-Reference

Consciousness

The nature of consciousness is difficult or impossible to explain, though everybody is sure he or she has it. Its study is a current "hot topic" that is approached from various directions. In earlier work, special attention was given to self-reference, the aspect of consciousness that allows an intelligent entity to be aware of itself. It is one of a number of characteristics that have been advanced as criteria distinguishing living from machine intelligence.

There are obvious ways in which consciousness has survival value, as it allows use of an animal or person's own intelligence as a model of that of others. Modelling is likely to be most accurate and therefore useful when it is of another member of the same species, but creatures have enough in common that interspecies modelling can also be valuable. A hunter, such as a cat, may to some extent use itself as a model to help guess the next movement of a mouse, and the mouse may use itself as a model of the cat.

People undoubtedly use consciousness in this way and are often explicit about it in saying something like: "If I were in his shoes, I would do such-and-such". It was suggested by Heinz von Foerster (von Foerster and Poerksen 2002) that the message conveyed by a talk is constructed by its hearers, who receive from the speaker only a series of clicks and grunts that constitute speech. This is in support of a general viewpoint termed "constructivism" that rightly denies the possibility of objective observation. However, the audience has more guidance in its choice of construct than the description suggests, as the series of clicks and grunts were devised by the speaker with the intention of producing a particular effect in his hearers' minds, using his own intellect as a model. Human, and probably much animal, communication must rely heavily on such modelling.

Hierarchies

Self-reference has aroused particular interest because it closes a loop and constitutes a form of recursion. It is also useful to consider intermediate levels of control, or at

A.M. Andrew, *A Missing Link in Cybernetics*, IFSR International Series on Systems Science and Engineering 26, DOI 10.1007/978-0-387-75164-1_6,
© Springer Science+Business Media, LLC 2009

least influence, within an intelligent entity. A number of puzzling aspects of intellect can be accounted for on the basis of hierarchical operation.

In digital computing, a hierarchical structure can be set up purely by programming, because a program can include subroutines that can be called from the main program flow, and they in turn call others and so on. This can be realised using information storage, or "memory" that is physically at a single level. However, it is pretty certain that the brain is not arranged in this way and that its hierarchy is "wired in". For one thing, the organization typified by the digital computer would be unlikely to have the tolerance of local damage shown by the brain.

Nevertheless, the running of programs on a digital computer illustrates one simple way in which a hierarchy can be useful. A program is an example of a finite automaton, equivalent to a Turing machine except in having finite memory, a distinction that is not significant in view of the enormous storage capacity of modern computers. It has been shown that there is no general way of determining whether a Turing machine running a particular program will halt (Minsky 1967; Copeland 2004). A number of results concerning computability follow from this. However, in everyday computing, no serious problem arises because the running of a program is supervised by a person or an operating system that terminates it when a certain time has elapsed, or patience expires, and fairly certainly biological intelligences have a similar safeguard against remaining in endless loops.

Amittedly, the practical computing situation differs from that visualised in treatments of the "halting problem" because the programs submitted are written with some purpose in mind and are meant to terminate within a time shorter than that allowed by the supervising person or system. Another significant aspect is that, in the days when computers were larger and slower and scarcer than now, an operator could watch a set of lights whose flashing indicated activity of parts of the machine, perhaps with a loudspeaker also coupled in, and it was possible to get a feeling for when the current program had gone wrong and possibly entered an endless loop. Although the detailed working of the brain is unlike that of a digital computer, it can usefully embody a supervisory system of this sort. A more powerful case for it will be made later, in discussing paradox. In both the brain and in computer administration, the supervisory mechanism has the character of a heuristic and does not need to be infallible in order to be useful.

(Strictly, as related by Copeland, the term "halting problem" was not introduced by Turing but by Martin Davis. However, the underlying theory is due to Turing, and it is usual to attribute the "halting problem for automata" to him.)

Description of the situation as hierarchical is necessarily with regard to a particular function or purpose, and many hierarchies may coexist in one system. For example, the termination of programs after a certain time, though also valuable in identifying faulty programs, is primarily a means of regulating access to the computing facility. It can be seen as part of a hierarchy in which the computer manager is above the users. On the other hand, some of these users may be running projects on which the prestige and viability of the entire organisation, including the computer unit, depend, in which case there is certainly a valid hierarchical description showing them above the computer manager.

Usually, the indication that something has gone wrong in the running of a program will be a repeating pattern of activity, and unproductive forms of mental activity have similar character. There is apparent correspondence to the supervisory control that has been suggested as facilitating adaptation or learning in continuous environments, where the responsiveness or loop gain should be as high as is compatible with stability. For operation in unknown environments, there is no general way of setting the gain suitably, except to let it increase until there is evidence of instability, whereupon it must be slightly reduced. Control allowing this would be hierarchical and could be a genetic forerunner of what is suggested above as playing a part in mental processes, and paralleled in administration of digital computers. This supports a view of intelligence as fractal, discussed in the next chapter. The need for supervisory control in continuous adaptation was referred to in connection with Fig. 5.2 and earlier in the discussion following Eq. (4.19) in Chapter 4.

Everyday Self-Reference

It is fairly obvious, and stressed by Sigmund Freud among others, that human thought operates on at least the two levels of the conscious and the subconscious. Even without invoking the latter it can be seen that a thought process at one level may monitor and partially control one at another. An example is the "crimestop" associated with "big brother" government in the book *Nineteen Eighty-Four* by Orwell (1989), according to which citizens were trained to detect that their thoughts were moving in a particular direction and to resist the movement. Something of the sort is apparent in the attitude of many people to religion, where it is rather usual to treat religion with respect, up to a point, but nevertheless to be sceptical of anyone who takes it really seriously. The ambiguity is captured in an item of graffity encountered in a public toilet:

When I talk to God it's called prayer
When God talks to me it's called schizophrenia

A similar attitude is implicit in a comment by Wiener (1950) where he compares the Roman Catholic Church with a communist regime. He notes that there is nothing a modern bishop wants less than a true saint arising in his diocese, and a commissar in a communist state has similar feelings about the possibility of a true revolutionary in his domain.

Self-awareness is also apparent in the desire of humans for self-respect, perhaps expressed as a desire for dignity or prestige. That the characteristic is not peculiar to humans is apparent from animal behaviour, for instance of domestic cats. Another form of self-awareness is represented by the "body image", or the picture a person or animal has of the boundaries of his or her physical embodiment (Nathan 1969). The acceptance of a limb as "belonging" seems to depend on its having a functioning sensory connection. A strange manifestation of the body image is the occurrence of "phantom limb" sensation after amputation, which according to Melzack and Wall

(1988) can be such that the amputee continues to be aware of a particular irritation, such as a thorn under a nail, in the limb no longer attached.

Yet more evidence for different levels of processing comes from the warm glow of satisfaction that comes when working on a mathematical or other problem, when the solution seems to be within reach though not actually achieved. In many contexts it seems appropriate to talk of "gut reaction" as a way of making a judgement when conscious reasoning fails to do so. These essentially emotional responses are at a different level from conscious thought, and conscious thought is also influenced by moods. An example is the generally recognised phenomenon of "wishful thinking" that can be seen as an opposite to the "crimestop" evasion of unwanted ideas.

Nontrivial Machines and Paradox

Von Foerster (von Foerster and Poerksen 2002) has described people as "nontrivial machines", where a trivial machine is one whose behaviour is readily predictable. A possible illustration, not from the book but in the spirit of it, would be comparison of the responses of trivial and nontrivial machines to the paradoxical statement "This sentence is false". A machine to be classed as trivial can be imagined cycling endlessly through the loop "if true, then false, so then true", and so on. A person remains in such a loop for only one or a very few cycles before recognising the situation as an endless loop or paradox, and perhaps naming it as one or the other. Philosophers have described a variety of paradoxes to illustrate diverse arguments, but perhaps the most significant aspect, common to all of them, is that a person can recognise a paradox as such.

If part of the thought process is considered to be a trivial machine that, by itself, would enter an endless loop, the fact that a paradox can be recognised and given exceptional treatment is consistent with supervision at a higher level. A connection can be seen with the insistence of Gordon Pask (1962), also discussed in Chapter 2, that learning should be considered as happening only when a new language is introduced. The word "paradox" or something equivalent, not necessarily verbal, is part of a new language of description. There is also a very suggestive connection with a phenomenon reported in connection with the "double bind" theory of Gregory Bateson, discussed below.

The recognition of paradox is thus something else that can be associated with supervisory control sensitive to cyclic behaviour, as considered previously and suggested to be an evolutionary development from continuous instability-detection. Very elaborate paradoxes may escape detection because of the length of the cycle, and complicated paradoxes-within-paradoxes (Scriven 1951; Gardner 1963) might also be missed, but these are exceptional. It is of course a common experience to be "tossed on the horns of a dilemma" when making difficult decisions in everyday life. This is cyclic behaviour, except that the supervisory mechanism of increasing desperation usually allows eventual escape, perhaps by the expedient of deciding to toss a coin.

(Gardner describes a "hanged man paradox" as a story about a man condemned to be hanged and told that the execution will be during a particular week. He was also promised by the judge that the arrival of the hangman on the day would be a surprise. The man started to laugh and claimed he could not be hanged according to this promise. He reasoned that, if the last day of the week had been reached, the arrival of the hangman would be no surprise so he could not be executed then. Also, once it is known that the last day is impossible, the same argument rules out the second-last day, which in turn rules out the third last, and so on. Unfortunately for him, however, once he had decided that he could not be hanged he was vulnerable, and when the hangman arrived on an arbitrary day, for instance the Wednesday, the prisoner was indeed surprised in accordance with the judge's promise. There are other versions of the story that refer to events less drastic than hanging, such as a class being set an examination.)

Gödel's Incompleteness Theorem

This famous theorem (Nagel and Newman 1959; Lucas 1961; Whiteley 1962; George 1962, 1972; Andrew 1979) states that in any consistent system that includes arithmetic there are formulae that cannot be proved within the system but can be seen by people to be true. This has sometimes been advanced as evidence that a Mechanist view of intelligence (that there is no essential difference between minds and machines) must be rejected.

The proof of the theorem as presented by Nagel and Newman depends on construction of a formula G with a numerical representation h referred to as its Gödel number. G and h are numerical representations of the assertion: "The formula with Gödel number h is not provable". The formula cannot be proved within an arithmetic system but can be seen by a person to present a paradox for which the only possible resolution is acceptance of the assertion as true. It is easy to feel that this is a thin argument for rejecting the Mechanist viewpoint, especially in view of the possibility of detection of paradox in a hierarchical system. The situation is in fact summed up in a remark of Donald MacKay, quoted by George (1962): "If you introduce the person (or mind) who goes beyond the deductive system, there is no reason why mind and deductive system should not be collectively mechanised".

Induction and Deduction

It is also easy to feel that arguments based on Gödel's theorem and Turing computability are unlikely to be relevant to the evolutionary origins of intelligence, which occurred before metamathematics, or formal mathematics of any kind, existed. George (1972) arrived at a similar conclusion by observing that living systems generally use inductive rather than deductive logic, and absolute consistency is unnecessary for the former. Most information processing in living systems is meant to produce some effect in the real world and consequently relies on the logically indefensible assumption that the future will be something like the past. These considerations will be discussed again in Chapter 8.

Double Bind and Creativity

The term "double bind" was used by Gregory Bateson (1973) to denote the situation where powerful motivations conflict. This may be "schizogenic" or productive of schizophrenia, and it was in connection with his work as a psychiatrist that he introduced the term. At the same time, however, at many points in his writings he associates the double-bind situation with creativity and mentions humor, art, and poetry.

In support of his contention, and also showing that the effect is not restricted to human psychology, he refers to a series of experiments on a dolphin. These took place in an exhibition tank and demonstrated "operant conditioning" to an audience. When the dolphin performed a specific act (in the first example raising its head), it was rewarded, first by hearing the trainer's whistle and then by receiving food. Only a few repetitions were needed to establish the pattern.

However, in the next session the dolphin was not rewarded for raising its head, but instead for a different specific act, namely a tail flip. Then in a third session the act that was rewarded was different again, and so on. The very interesting observation is that when the dolphin came "on stage" for the fifteenth session she put on an elaborate performance including eight conspicuous pieces of behaviour of which four were entirely new. Not only were they new for this particular dolphin but also had never been previously observed in the species.

The account is in terms of discrete events, responses, and rewards, and therefore not immediately applicable to the central problem considered here, namely the transition from continuous to discrete processing. However, as was argued for the conditioned reflex, operant conditioning could be extended to include learning of continuous skills, and the emergence of a double bind would then correspond roughly to the detection of a need to classify environments, as discussed in Chapter 4. The parallel ends there, because the creative response of the dolphin, in inventing new behaviour patterns, is of a different nature to the hypothesised creative response of devising categories. Nevertheless, the comparison strongly suggests a common underlying mechanism.

Summary of Chapter 6

In order to consider ways in which thought can be considered to be hierarchical, aspects of consciousness are reviewed. A functional view is taken, with consciousness seen as arising as a means of using the conscious entity's self as a model of other creatures with which it interacts. This can apply for example to predator–prey interaction and also to interpersonal communication.

It is useful to consider levels of control, or at least influence, in an intelligent system. For robust operation, this should be wired-in rather than depending on stored addresses as in computer programs. An important function of higher-level or supervisory control must be the detection of repetitive or cyclic behaviour which may indicate encountering a paradox. This can explain the ability to recognise a paradox as such and to avoid being trapped by it as might happen to a "trivial machine" as postulated by von Foerster. Importantly for the theme of the book, a connection is indicated between this type of supervisory control and the detection of incipient instability that is necessary for continuous-variable adaptation.

That human intelligence is hierarchical is illustrated in a number of familiar contexts including the "crimestop" postulated by Orwell in *Nineteen Eighty-Four*. Other examples are the coexistence of the conscious and subconscious mind, and the body image referred to by neurologists and evident in "phantom-limb" sensations.

A connection is suggested with Gregory Bateson's principle of "double bind" and its association with creativity. The "double bind" is similar to paradox, and the recognition and perhaps naming of a paradox can be a form of creativity. Observations on a dolphin lend support to the viewpoint.

Chapter 7
Fractal Intelligence

Is Intelligence Fractal

In the Past chapter, it was suggested that there is a connection between the detection of cyclic activity at the conscious level, perhaps arising from paradox or dilemma, and the detection of incipient instability at the continuous-control level. In each case, detection is normally followed by action that suppresses the cyclic activity. It is further suggested that in biological systems, the continuous-control action should be seen as the more primitive, with conceptual-level behaviour evolved from it but not displacing it.

The acquisition of skill in tasks like riding a bicycle, or acting as a self-stabilising servo in various situations, is readily termed "learning", but there is a tendency to think of it as distinct from learning at the intellectual or discrete-logic level. This may be partly because, if all the physical aspects (masses, elasticities, viscosities, and physical arrangement) are known, mathematical analysis of a continuous task may be possible. However, people and animals excel at learning continuous tasks without formal analysis and where essential parameters are not known. This implies learning how the environment behaves, equivalent to forming a model of it. Learning may be wholly or partly by imitation, corresponding to rote-learning in linguistic contexts.

Elementary Exemplification

A number of important ideas connected with AI have an "elementary exemplification" (Andrew 1989, 1992) in terms of continuous variables. For example, Minsky's "heuristic connection", introduced by him with reference to complex problem-solving, has an elementary exemplification in the exploitation of environment continuity by process controllers and by people and animals performing acts of skill. The similarity of the problem of what to do when a bicycle has tilted to 3 degrees and what to do for 4 degrees is an elementary exemplification of heuristic connection between problems.

A.M. Andrew, *A Missing Link in Cybernetics*, IFSR International Series on Systems Science and Engineering 26, DOI 10.1007/978-0-387-75164-1_7,
© Springer Science+Business Media, LLC 2009

Another clear example is the "means-ends analysis" that is a feature of the General Problem Solver (GPS) of Newell *et al.* (1959), which amounts to repeated asking of the question "What is the difference between what I have and what I want?" followed by "What action would reduce the difference?" A simple feedback controller such as a room thermostat can be seen as an elementary example of this, but of course the kinds of difference identified by the GPS when applied to theorem proving in mathematics are much more subtle and varied in nature than is, for example, the deviation of room temperature from a set point.

In the introduction to his famous book, Wiener (1948) talks about negative-feedback control of the task of picking up a pencil, even though no simple numerical feedback can be used. The feedback needed must include something like the means–ends analysis of the GPS though at the same time it has continuous-variable aspects. The example illustrates the intimate mixing of continuity and discrete-event operation that is characteristic of much biological activity.

Minsky and Selfridge (1961) acknowledge the possibility of treating elementary and complex measures in similar terms in discussing their "basic learning heuristic". This amounts to the apparently simple rule: "In a novel situation try methods like those that have worked best in similar situations". At first sight this seems to be so obvious as to be hardly worth stating, but two kinds of undefined similarity are involved, namely similarity of situation and similarity of method, and as these authors say:

> . . . advanced forms [of learning] which discover and use new categories can cover the creative aspects of genius.

The reference to new categories implies new kinds of similarity, and these authors effectively acknowledge that a rule whose "elementary exemplification" seems simple and obvious is applicable in ways that are far from trivial.

It has to be admitted that the view of biological evolution as proceeding from continuous control to concept-based processing is not universally accepted, and certainly the emergence of a digital genetic code plays an additional part. The view that continuous control is an appropriate starting point is supported by Churchland (1986) and by a comment attributed (Andrew 1987) to Sherrington, the "Father of Neurophysiology", to the effect that the brain should be seen as a development of motor control.

It is interesting to note, however, that Gabor did not consider the operation of his learning filter to be relevant to the brain. In the discussion appended to the paper (Gabor *et al.* 1961), in reply to a question from Gordon Pask, the authors say that, despite their use of the word "learning", they regard the filter as having no relation to the brain. They claim that the quantitative operation of the filter has no correspondence to natural intelligence, which they describe as classificatory. Such a view of brain function is widely held and implicit in much work in mainstream AI, but argued here to be an unsuitable starting point for its consideration.

It often happens that a complex phenomenon is described as though it was identical with its elementary exemplification, often without the person being aware of the conceptual jump involved. For example, the idea of distance from a solution in problem solving (MacKay 1959) seems remarkably natural, even when no means of computing a numerical criterion has been devised.

It is also easy to believe that the principle of elementary exemplification has a bearing on the backpropagation of errors, alternatively termed "significance feedback" with an odd objective function (Andrew 1993), as applied in practical applications of artificial neural nets. Current applications operate in terms of simple numerical error, but introspection suggests that an analogous process is a part of higher-level learning and conscious thought, where "errors" are likely to be expressed in terms of structured variables, as when a person says "I should have been more forceful" or "more tactful", and so forth, and the adaptation that will affect future behaviour is correspondingly structured. Backpropagation as currently implemented could be seen as an elementary exemplification of the structured version. Just how the structured version might be realised in neural nets is difficult to say, but seems a fruitful area for speculation.

Fractal Intelligence

The ideas in this chapter suggest a possible reason for the failure of known evolutionary or self-organising principles, embodied for example in artificial neural nets, to achieve the required transition from continuous to concept-based processing. It may be necessary for the elementary (pretransition) organisation to play dual roles. One is to provide the initial state, or substrate, from which the new one develops, but it must also provide a kind of blueprint of which the pretransition version might be seen as an elementary exemplification. It is certainly not easy to see how the "blueprint" function could be implemented, but the reappearance of elementary principles in the higher-level context suggests it plays a part and that the want of such a function may be a reason for the failure of self-organising networks to achieve the transition.

The dual role of the pretransition organisation in providing both substrate and exemplar invites comparison with the emergence of recursive or fractal structures in the physical form of organisms. As is clear from illustrations in Peitgen *et al.* (1992), fractal growth produces lifelike forms, suggesting a correspondence to natural development. It is reasonable to suppose that intelligence may have analogous characteristics.

Summary of Chapter 7

It is shown that a number of principles associated with mainstream AI, and expressed in terms of discrete concepts, have "elementary exemplification" in terms of continuous variables. It is suggested that this observation may hold the key to a means of usefully extending the field of application of backpropagation of error in neural nets.

It is also suggested that the "elementary exemplification" principle may indicate an important characteristic of the evolution of intelligence, namely that the more primitive state from which evolution stems must serve two functions, denoted by substrate and blueprint. That intelligence should have the fractal character that this— and the general idea of elementary exemplification—implies is in keeping with other evidence that fractal growth simulates living development.

Chapter 8
Conclusions

Motivation

What has been presented is an idiosyncratic view of cybernetics and artificial intelligence. It stems from very early interest in modelling nervous activity, when most modelling attempts assumed the all-or-nothing "logical" behaviour of McCulloch–Pitts model neurons. The then-young Artificial Intelligence approach also favoured discrete logic.

The need to exploit continuity became clear when an application to control of an industrial plant was considered, as discussed in Chapter 3. The number of discriminable states of a typical plant is enormous, and to associate a separate learning task with each is uneconomical and unfeasible. The learned responses must somehow be smeared over states that are metrically close. The need for such smearing was illustrated in Chapter 3 by reference to a person learning to ride a bicycle.

As shown in Chapter 4, the result can be achieved by arranging that the control signals to the plant are computed as functions of sensory data from the plant, and in an adaptive or learning controller the functions are modified by experience. This takes account of some indicator of success, or "hedony", and adjustment is of parameter values, and also of the form of the computed function. Plans for self-adjusting controllers of this kind were discussed by Andrew (1959, 1961) and independently by Gabor *et al.* (1961) as well as in a great deal of other work associated with control engineering, and by Samuel (1963) in the means used to adjust the scoring polynomial of his checkers-playing program.

Because neurons can act as continuous transducers, a self-adjusting continuous controller is readily formed as a neural net, and many practical applications of artificial neural nets depend on exactly this, usually with adjustment by backpropagation. This is neural modelling on a different basis from that directly inspired by the treatment by McCulloch and Pitts, and it is important to consider the connection between them and their relevance to real neural nets.

Another interesting connection is between consideration of continuity in these terms and the idea of "heuristic connection" between problems introduced by Marvin Minsky, first in discussion of the paper by MacKay (1959) and then in Minsky (1963). This provides an interesting link with the more "intellectual" types

A.M. Andrew, *A Missing Link in Cybernetics*, IFSR International Series on Systems Science and Engineering 26, DOI 10.1007/978-0-387-75164-1_8,
© Springer Science+Business Media, LLC 2009

of problem dealt with in mainstream AI. There is here a particularly clear case of an "elementary exemplification", where the complex or nonelementary example is the perception of heuristic connection in mainstream AI, and the elementary version is the direct metrical similarity of situations of the bicycle-riding kind (Andrew 1992c).

The contents of the current book result from consideration of these connections over a great many years. The treatment of adaptive continuous control in Chapter 4 is now one contribution among many to the topic but is at a level that may correspond to its appearance in biological evolution. It also illustrates in a particular context the equivalence of methods that do and do not embody a model. The enhancement to take account of states of the environment, due to Sutton and Barto (1998), is clearly important but probably corresponds to a later stage of evolution.

The need to acknowledge and exploit continuity has also been appreciated by Zadeh (1965) whose fuzzy set theory has been the basis of a vast amount of theoretical work with important practical applications. As discussed in Chapter 3, however, this development cannot correspond to biological evolution, where continuous processing certainly came before spoken language. Correspondence to biological evolution may not be absolutely necessary for the achievement of powerful artificial intelligence, but it is difficult to imagine it being done otherwise.

Is Artificial Intelligence Possible?

The discussion has been based on what seemed to be a widely accepted view at the time of inauguration of cybernetics, namely that the brain is an information processor and that there is no reason except complexity, and of course the difficulty of finding how to do it, that its function should not be copied in artifacts. This is one formulation of the "strong AI position" of which a variety of alternative formulations are reviewed by Bishop (2002). To an extent that I find surprising, the viewpoint is currently contested. In the early days the debate was mainly about machine capabilities in various set tasks, such as playing a strong game of chess, but attention has been given increasingly to the question of machine "mind" or understanding or consciousness. This of course raises philosophical issues about the nature of consciousness in general. Notable contributions opposing "strong AI" are those of Searle (1984) already mentioned and the formidable treatment by Penrose (1989). A simple observation that also goes against it is that although a machine can play a powerful game of, say, chess, it is difficult to imagine that it feels elation on winning or disappointment on losing, in ways corresponding to human emotions. It can easily be programmed to produce an exclamation of triumph on winning or a message of congratulation to the opponent on losing, but these cannot carry much conviction as evidence of true emotion.

The paper of Alan Turing (1950) shows remarkable imagination and foresight, considering that it was written when electronic computers were in their infancy and large-scale calculations were performed by other means. His arguments remain substantially valid today. His suggested criterion of computer intelligence is the "Imitation Game", requiring that a person interacting with, sometimes, a computer, and sometimes a person, should be unable to tell which is which. It is suggested that communication should be through a teleprinter link or an intermediary, so that such things as visual appearance and tone of voice are not relevant. It would also be necessary to have the computer delay some of its responses, especially to questions involving calculations, and also to introduce errors, but both of these could be easily arranged.

The relevance of the Imitation Game has been contested on various grounds, and it is true that in situations where the range of topics was restricted, the test might be said to have been passed, rather trivially. People interacting with a number of AI programs have been convinced they were communicating with a person, but communicating in a restricted way. The most notable example is the ELIZA program devised by Weizenbaum (1976), also discussed by Boden (1977). This program was designed to simulate a psychiatrist of a particular school, and some of the patients "interviewed" were convinced they had been in communication with a real doctor. This was possible because the simulation was of nondirective psychotherapy in which the therapist avoids "priming" the subject with any morsel of information. The operation of ELIZA depends on a set of fairly simple tricks. For example, statements by the patient beginning with "Can I . . ." or "Should I . . ." or "I would like to . . ." are subjected to some fairly simple syntactic manipulation and come back as "Why do you want to . . ." and so on. Another trick is to note the occurrence of some significant words such as family relationships, and if for example "father" has been mentioned several times the pseudo-psychiatrist may prompt: "Tell me about your father".

Interaction with a simulation of a nondirective psychotherapist is of course a very different matter from unrestricted interaction. A machine that gave a convincing performance in the latter would surely have to be recognised as intelligent, because it is by similar interactions, possibly but not necessarily supplemented by observations of unrelated behaviour, that we judge the intelligence of fellow humans. Success in the test must in fact be a sufficient but not a necessary condition for intelligence, as for example animals can be seen as intelligent although they do not use human-type language.

Arguments of the type of Gödel's theorem are mentioned cursorily by Turing. As argued in Chapter 6, such considerations cannot have any bearing on the evolutionary origins of intelligence. Implications of the theorem are also discussed in an early influential work by Arbib (1964). He explains the importance of the theorem by reference to a Formalist view of mathematics due to Hilbert, shown by the theorem to be untenable. He proves the theorem in a way that is more formal than that of Nagel and Newman (1959) and of Penrose. Arbib is emphatic in saying that the theorem gives no grounds for doubting that machines can be intelligent.

The "strong AI position" is vigorously opposed by Searle (1984) using the "Chinese Room" argument, and further views prompted by this are the subject matter of a collection by Preston and Bishop (2002). Searle likens a computer to a person, ignorant of the Chinese language, sitting in a room and receiving inputs in

Chinese. He supposes the person to be able to recognise and classify the inputs and to consult a large rule book that indicates appropriate responses, also in Chinese. It is argued that the overall effect, with an adequate rule book, could give the appearance of the system understanding Chinese although the operator, and therefore the computer he models, remain ignorant of it.

Because the constructions possible in a natural language have infinite variety, the rule book needed to let such a system pass Turing's test, applied in Chinese, would be enormous, and in fact impossible if the test were applied sufficiently assiduously. Searle's contention is that human appreciation of semantic content is impossible for machines. It is likely to appeal on essentially emotional grounds and to that extent can be attributed to conceit on the part of humans. Searle also refers however to the enormous background, first of evolution and then of life experience, determining human responses, and this is certainly a significant consideration. The earlier observation, that a present-day chess-playing computer would not feel elation or disappointment, might be attributed also to want of background. It is easy to imagine an evolving species of animal or machine coming to incorporate a measure of confidence (versus diffidence) as a generalisation of its recent history of successes and failures, and feelings of triumph or disappointment would correspond to adjustments of this.

The emotional desire to accept Searle's contention has something in common with the view, encouraged by religions but now seen as dangerous by environmentalists, that humans occupy a privileged and protected place in the scheme of things. What may be seen as another warning against undue reliance on impressions is an amusing account of the refusal of the poet Goethe, who expected to be remembered as a scientist rather than as a poet, to accept that such a pure elemental thing as white light could possibly be broken down into the colours of the spectrum, or formed by their combination. He even accused Newton of falsifying results. The account is by Sherrington (1949), the "Father of Neurophysiology", though with no direct bearing on that topic.

Searle's argument appears at first sight to defend a dualist view of nature, such that living matter, able to react to semantic content, is fundamentally different from nonliving. Surprisingly (to me at least) this is not his intention, and he blames the AI movement for introducing dualism in the suggestion that intelligence could reside in a machine. His defence of his stance by reference to the evolutionary and experiential background of biological intelligence seems to make his argument one of complexity rather than of principle (though terms like "principle" and "essence" in such contexts are admittedly imprecise). Reference to monism and dualism can only refer to the physical media (hardware and wetware) on which the systems are imposed, and there is clearly something additional to this in organisation, represented in the computational case mainly by the program.

Even with reference only to the hardware and wetware, the distinction between monism and dualism is much less clear-cut than it seems at first. At the time of my own first acquaintance with neurophysiology, the monist view seemed to be adequately defined by saying that the brain worked according to principles of physics and chemistry. This was at a time when, although relativity and quantum theory were well established, the tendency was to think of physics in Newtonian terms. In more recent time it is much less clear just what is implied by conformity to laws of physics, as Penrose makes clear.

Penrose (1989) opposes the "strong AI position" using formidable arguments based on advanced mathematics and quantum theory. He suggests that phenomena of consciousness are nonalgorithmic and so cannot be imitated by a computer. Quantum theory connects with consciousness in at least two ways. One refers to the contrast between the quantum-mechanical view of particles as waves that can for example pass simultaneously through two slits of a spectroscope, and the alternative view supported by other phenomena in which "particles" behave as such. There is a connection with consciousness because it seems to be the act of observation, or of detection by say a photocell, that forces the particle to have a unique location. Consequences of this are discussed with reference to Schrödinger's unfortunate, but fortunately hypothetical, cat, whose fate, inside a sealed enclosure, is dependent on an indeterminate event of particle detection. That detection is necessarily indeterminate follows from the fact that the waves represented by Schrödinger's equation are interpreted as indicating the probability of the particle being at any specified point.

The thought experiment suggested by Schrödinger requires imagination of a cat in a sealed enclosure, allowing no communication in or out except means whereby detection of a particle (by a photocell or Geiger counter) may cause the breaking of a phial of cyanide inside the enclosure and the death of the cat. The original suggestion by Schrödinger (Moore 1989) was that the event would be triggered by emission of a particle from a quantity of radioactive material such that the probability of emission in the time of exposure was 0.5, and Penrose refers instead to the passing of a single photon through a half-silvered mirror. The point at issue is the state of the cat during the time between its possible execution and the opening of the chamber for observation. Some workers, though not Schrödinger or Penrose, have claimed that the cat is neither dead nor alive, or in some way combines both states, until it is observed.

The implications of this thought experiment for human consciousness are certainly not obvious, but Penrose makes another observation that could have a direct bearing on neurophysiology. Under some circumstances, quantum mechanics allows results that seem to defy explanation in terms of algorithms. It is as though all possible configurations of a situation were generated and one chosen to fit a set of boundary conditions, a result that could in principle be achieved by algorithmic means but only at enormous and probably prohibitive computational cost. The effect is most readily explained, and in fact suggests, the "many-worlds" or "parallel-universe" interpretation of quantum theory, which however meets with difficulties elsewhere. Penrose describes observations on crystal formation in which atoms form configurations compatible with their interactions, but which could not have been reached by successive addition of atoms, which seems to indicate a "many-worlds" quantum effect. He makes the plausible suggestion that something similar may occur in the adaptive modification of adjacent synapses on a neuron.

Various results bearing on capabilities of computers depend on representation of the computer as a finite automaton, whose states can be listed and correspond to rows of a table showing outputs and transitions, and a similar view is implicit in Searle's representation as the Chinese Room. The validity of such representation is undeniable, but the number of states readily exceeds Eddington's estimate of the number of electrons in the universe, or even that number raised to a high power. (The number is 1.29×10^{79}, or about 2^{263}, so is the number of states of a store of only 263 binary digits.) Practical computer programming has a quite different "feel", with the possibility of modelling any specified physical system with arbitrary precision. This gives confidence that at least a significant part of brain function will also yield.

One way that the "strong-AI position" has been epitomized is in a claim that there is no essential difference, except complexity, between a room thermostat "knowing" that a room is warm or cold and a person making the same discrimination. It is interesting to examine further the human response to temperature. As discussed by Cornew et al. (1967), the core temperature of the body is controlled very precisely, in two stages. The temperature of the periphery is regulated relatively crudely by a variety of mechanisms, including shivering and sweating and also including many behavioral responses such as seeking favourable locations, donning clothing, building houses, and so on. The core temperature is regulated more precisely by control of blood flow in vessels linking the core to the periphery. The final critical control is thus achieved by a means that is similar to the room thermostat without involvement of consciousness.

The contrast between the two mechanisms suggests that evolution invokes consciousness as a means to an end. For fine control of core temperature, vasoconstriction and dilatation provide a direct means that can operate without consciousness, but regulation at the periphery has to be largely behavioural. An interesting point is that the migration of humans to almost all parts of the earth's land area has been vastly accelerated by behavioural stratagems, particularly that of wearing clothes. The migration of other species, such as of bears into the Arctic, had to depend on the slow evolution of the necessary fur and so on, but humans have been able to colonize diverse areas with almost no biological adaptation. In the Arctic this has been partly by exploiting the adaptation of the bears and using their fur for clothing, but the point is that humans achieve the result by essentially behavioural means, depending on conscious perception.

This utilitarian view of consciousness receives further support from the fact that immobile life forms, plants and fungi, do not seem to have consciousness, presumably because they do not need the complex communication and modelling that it would allow. Penrose mentions the modelling aspect, with reference to its usefulness to a predator in predicting the movements of its prey, as suggested earlier here with reference to a cat and mouse. He rejects the suggestion of an evolutionary origin for consciousness due to something like this, but on what seem to me to be inadequate grounds. He argues that a conscious predator could only use itself as a model if the prey was also conscious, and so the problem of identifying an origin for consciousness has something of the character of the well-known query as to which came first, the chicken or the egg.

This argument seems to be flawed in two respects. In the first place, a conscious entity can perfectly well use a part of itself as a model of something not necessarily conscious, for instance when the function of a room thermostat is most readily explained by observing that without it the heating system would not know whether the room was hot or cold. Such an explanation arises readily but does not imply that the heating system is thought to be alive, let alone conscious. Another point is that modelling of the kind visualised can be useful within a single species, perhaps in allowing communication, and fairly certainly when its members are in combat. Within a single species consciousness could emerge gradually. Combat between members of the same species, especially between males competing for females, is common, and the fact that it often ends in intimidation without serious injury tends to support the involvement of modelling.

The emphasis on utility need not exclude the possible emergence of finer feelings of kindness and compassion. Ways in which altruism could evolve have been considered by Haldane (1955), Hamilton (1964), Ozinga (1981), and notably by Axelrod (1984).

Among the pioneers of cybernetics, Heinz von Foerster (von Foerster and Poerksen 2002) left no room for doubt that he opposed the "strong AI" viewpoint and insisted that AI programs should be seen merely as metaphors for brain activity and that brains belong to a class of systems that by their nature cannot be analysed. On the other hand, the enthusiasm of Norbert Wiener (1948) for the correspondence between a circuit diagram for a pattern-recognising artifact and the neural connections of a layer of the cerebral cortex suggests he would support "strong AI". The readiness of Ashby (1956, 1960) to represent nervous and biological systems as state-determined suggests he would concur. It can also be assumed that Warren McCulloch would have supported the viewpoint, because as related in Chapter 1, he hoped to advance his lifetime quest of understanding human understanding from the results of electrophysiological recording. Marvin Minsky and other early workers in AI were much influenced by McCulloch.

Probably Academic

However, this speculation about "strong AI" and the ultimate possibilities of machine intelligence is fairly certainly academic as far as anything in the reasonably foreseeable future is concerned. My own feeling is that there is no definite limit to what is possible, but the difficulties are enormous. Disappointment with mainstream AI was mentioned in Chapter 1.

The early construction of a "tortoise" or *machina speculatrix* by Grey Walter (1953) was claimed to demonstrate that interesting behaviour, perhaps to be described as intelligent, could be achieved fairly simply. There is a similar implication in the description of the "Seven Dwarfs" set of robots by Kevin Warwick (1998). These schemes do show behaviour that is intriguingly and compellingly lifelike, and that may have some bearing on the early evolution of intelligence, but the suggestion that the working of the brain is simple is just wrong. At the end of a valuable review of aspects of machine and natural intelligence, Fischler and Firschein (1987) quote from another author:

> I suspect that the representational system with which we think, if that's the right way to describe it, is so rich that if you think up any form of symbolism at all, it probably plays some role in thinking.

A viewpoint, backed up by practical results, and very much in agreement with the message of this book is due to Brooks (1999). He has developed methods of robot control that differ from previous AI approaches in having much closer coupling of sensory inputs to motor outputs. Because the inputs and outputs are mostly continuous, this agrees with the arguments here, though presented with a different emphasis. The relevance of robotics is also part of the argument of Churchland (1986), and it is reasonable to think that considerations arising in this context should correspond to those of the natural evolution of intelligence. Brooks' methods are in contrast to much earlier work on AI as applied to robotics and to the theorising about human perception that went with it.

In the earlier work, it was assumed that the task of robot control could be broken down into perception, modelling, planning, task execution, and motor control as separate stages, and an enormous amount of effort was devoted without appreciation of the difficulty of coupling these to the real world. As related in some detail by Brooks (1989), the much-acclaimed robots referred to as Shakey and Flakey operated only in specially adapted static environments, and even then painfully slowly. Separate stages of modelling and planning are obviously unsuitable for operation in a changing environment.

The problems associated with achieving worthwhile artificial intelligence are such that it is necessary to look for all possible clues from the living example. It is argued by Brooks, as well as here, that the evolutionary development of intelligence is significant. An obvious difficulty is that the development happened long ago and we can only surmise the details, although, following the example of Darwin, we can make inferences by observing existing life forms. Brooks argues for the development of adaptive behaviour in stages, with each stage growing on a substrate of those before, and has used this principle effectively in robot design.

It is possible that intelligence as we know it can only be exhibited by something that has a long history of survival in a complex environment. As argued above, the distinction made by Searle in his Chinese Room argument can be construed in these terms. The difficulty then has to be seen as a matter of complexity rather than of principle because, if an intelligent machine is possible, there has to be a remote possibility that by some fantastic stroke of luck someone might accidentally construct it, though the probability is infinitesimal.

Another way of expressing much the same limitation is to say that the goals set to AI programs have little relevance to primitive survival. Early studies in mainstream AI were influenced by the work of Polya (1954) on inference in mathematics, and although this should be part of the "big picture" of intelligence as a whole, the goals that are set cannot have much relation to primitive evolution. Even the goals set to simple robots, like seeking sources of electrical energy when their batteries run low, or of collecting trash or land mines, have an artificial flavour, though as developed by Brooks with extension to map-building may give important insights.

The other way of seeking help from natural intelligence is to examine the internal working of the brain. The hope of copying this in artifacts, or at least of drawing useful parallels, was part of Wiener's reference to "the animal and the machine" and of McCulloch's hope of furthering his lifetime quest (epitomized as: "What is a man, that he may know a number . . .") by physiological experimentation, initially with mirror galvanometers.

Since then, a great amount of progress has been made in analysing the working of the brain, using tools enormously more sensitive and convenient than mirror galvanometers. Much has been learned about sensory pathways, especially that of vision and including representation in the cerebral cortex and other brain structures. A great deal is known about interconnections within the brain, partly as a result of a technique of strychnine neuronography of which McCulloch was a pioneer. Imaging techniques, especially nuclear magnetic resonance, have yielded much valuable information. They can be used to localise activity in the brain of a conscious subject while visual images or other stimuli, or instructions to perform mental tasks, are presented.

Despite all this, there are very fundamental aspects of brain activity that remain mysterious. The imaging techniques respond to metabolic effects of neural activity, so cannot reveal fine details. It is not even certain whether the brain is to be explained exclusively as a neural net. Some workers, notably Hameroff (see Dayhoff et al. 1993) and Penrose (1989, 2002) seem fairly sure that microtubules forming the cytoskeleton of cells have an important communication function and contribute to the computation. Cells other than neurons, but in close proximity to them, have abundant microtubules and may play a part. These fundamental aspects, as well as the possible part played by quantum mechanics, are still matters for speculation.

Also, there is no clear picture of the mechanisms underlying memory. That synaptic modification plays a part is confirmed by a vast amount of experimental data, but it does not account for the phenomena of either long-term or very short-term memory. The early work of von Foerster (1950) suggests a molecular mechanism for long-term memory, and there have been suggestions that something akin to the genetic code could be involved. Even if its basis were identified, unravelling the means of transfer between it and short-term memory would be a further challenge.

There is certainly a long way to go before artificial intelligence can be modelled on the natural variety in any detail, and it is extremely unlikely that Turing's test, without restriction of subject matter, will be passed in the reasonably foreseeable future. This is not to deny that progress can be made towards less ambitious goals, with attention to the interface between continuous and concept-based processing a neglected vital ingredient.

Summary of Chapter 8

Motivation for the approach giving rise to the book is explained in terms of an early study of machine learning, intended to be on a neural basis. Two questions that arise are (a) that of relating very different approaches to neuro-modelling, differing in the importance attached to continuity, and (b) that of relating simple metrical continuity with less tangible continuity of the "heuristic connection" kind. An alternative way of acknowledging the importance of continuity and treating it is by the introduction of fuzzy set theory and fuzzy methodology, but reasons are given for believing this is not the most fundamental approach.

Early treatments of cybernetics and AI make an often tacit assumption that virtually unlimited machine intelligence is possible, an assumption that has been challenged by, for example, the Chinese Room argument of Searle and the later work of Penrose. The "strong AI" view is defended here, with the observation that consciousness has a utilitarian aspect that may largely account for its evolution. It is acknowledged however that the current state of knowledge, especially of the intimate working of the brain, is such that speculation about ultimate possibilities of artificial intelligence is likely to remain academic for the reasonably foreseeable future.

References

Adler, J. (1976) "The sensing of chemicals by bacteria", *Scientific American*, vol. **234**, no. 4 (April), pp. 40–47

Alberts, B., Johnson, A., Lewis, J., Raff, M, Roberts, K. and Walte, P. (2002) *Molecular Biology of the Cell*, 4th edn., Garland, N.Y., pp. 890–892. (Also treated in earlier editions, by Alberts, B., Bray, D., Lewis, J., Raff, M, Roberts, K. and Watson, J.D., 1983, 1989, 1994, on pages indicated by "Bacterial Chemotaxis" in subject indices.)

Aleksander, I. and Morton, H. (1995) *An Introduction to Neural Computing*, 2nd edn. Thomson, Boston, pp. 115–133

Allison, A.C. and Nunn, J.F. (1968) "Effects of general anæsthetics on microtubules: A possible mechanism of anæsthesia", *The Lancet*, Dec. 21, pp. 1326–1329

Allison, A.C., Hulands, G.H., Nunn, J.F., Kitching, J.A. and Macdonald, A.C. (1970) "The effect of inhalational anaesthetics on the microtubular system in *actinosphaerium nucleophilum*" *J. Cell Sci.* **7**, pp. 483–499

Anderson, J.A. (1996) "From discrete to continuous and back again", in: Moreno-Diaz, R. and Mira-Mira, J. (ed) *Brain Processes, Theories and Models: An International Conference in Honor of W.S. McCulloch 25 years after His Death*, MIT Press, Cambridge, Mass., pp. 115–124

Andrew, A.M. (1959a) "The conditional probability computer", in: *Mechanisation of Thought Processes* (Proceedings of Conference in the National Physical Laboratory, Teddington), HMSO, London, pp. 945–946

Andrew, A.M. (1959b) "Learning machines", in: *Mechanisation of Thought Processes*, HMSO, London, pp. 473–509

Andrew, A.M. (1961) "Self-optimising control mechanisms and some principles for more advanced learning machines", in: *Automatic and Remote Control* (IFAC Moscow Congress), Butterworth, London, pp. 818–824

Andrew, A.M. (1965) *Significance Feedback in Neural Nets*, Report of Biological Computer Laboratory, Univ. of Illinois, Urbana

Andrew, A.M. (1967) "To model or not to model", *Kybernetik*, vol. **3**, no. 6, pp. 272–275

Andrew, A.M. (1973) "Significance feedback and redundancy reduction in self-organizing networks", in: Pichler, F. and Trappl, R. (ed) *Advances in Cybernetics and Systems Research 1*, Hemisphere, London, pp. 244–252

Andrew, A.M. (1978) "Succinct representation in neural nets and general systems", in: Klir, J.G. (ed) *Applied General Systems Research: Recent Developments and Trends*, Plenum, New York, pp. 553–561

Andrew, A.M. (1979) "Gödel, mechanism and paradox" (letter), *Kybernetes*, vol. **8**, no. 1, pp. 81–82

Andrew, A.M. (1981) "The concept of a concept", in: Lasker, G.E. (ed) *Applied Systems and Cybernetics*, Pergamon, N.Y., pp. 607–612

Andrew, A.M. (1987) "Self-organizing systems and Artificial Intelligence", *Systems Research and Info. Science*, vol. **2**, pp. 143–151

Andrew, A.M. (1988a) "Computing statistics as running values", *Systems Research and Info. Science*, vol. **2**, pp. 237–243

Andrew, A.M. (1988b) "Tesselation Daisyworld – a new model of global homeostasis", in: Trappl, R. (ed) *Cybernetics and Systems '88*, Kluwer, Dordrecht, pp. 313–320

Andrew A.M. (1989) *Self-Organizing Systems*, Gordon and Breach, London

Andrew, A.M. (1990) *Continuous Heuristics: The Prelinguistic Basis of Intelligence*, Ellis Horwood, Chichester

Andrew, A.M. (1991) "Continuity and Artificial Intelligence", *Kybernetes*, vol. **20**, no. 6, pp. 69–80

Andrew, A.M. (1992a) "Is the representation of uncertainty the main issue?", in: Trappl, R. (ed), *Cybernetics and Systems '92*, World Scientific, Singapore, pp. 41–48

Andrew, A.M. (1992b) "More thoughts on continuity and logic", *Kybernetes*, vol. **21**, no. 4, pp. 43–46

Andrew, A.M. (1992c) "Elementary continuity and Minsky's 'heuristic connection'" *Int. J. Systems Research and Information Science*, vol. **5**, pp. 147–154

Andrew, A.M. (1993a) "Real and artificial neural nets", *Kybernetes*, vol. **22**, no. 8, pp. 7–20

Andrew, A.M. (1993b) "Significance feedback in neural nets" *Int. J. Systems Research and Information Science*, vol. **6**, pp. 59–67 (Reprinted from report of Biological Computer Laboratory, University of Illinois, 1965)

Andrew, A.M. (1994a) "What do we mean by 'knowing'?", *Cybernetica* **3/4**, pp. 205–214

Andrew, A.M. (1994b) "What is there to know?", *Kybernetes*, vol. **23**, no. 6/7, pp. 104–110

Andrew, A.M. (1995a) "The decade of the brain – some comments", *Kybernetes*, vol. **24**, no. 7, pp. 25–34

Andrew, A.M. (1995b) "The decade of the brain – addendum and review", *Kybernetes*, vol. **24**, no. 8, pp. 54–57

Andrew, A.M. (1997) "The decade of the brain – further thoughts" *Kybernetes* vol. **26**, no. 3, pp. 255–264

Andrew, A.M. (2000) "Self-organisation in artificial neural nets", *Kybernetes*, vol. **29**, no. 5, pp. 638–650

Andrew, A.M. (2001a) "The poet of cybernetics", *Kybernetes*, vol. **30**, no. 5/6, pp. 522–525

Andrew, A.M. (2001b) "Backpropagation", *Kybernetes*, vol. **30**, no. 9/10, pp. 1110–1117

Arbib, M.A. (1964) *Brains, Machines and Mathematics*, McGraw-Hill, New York

Arbib, M.A. (1970) "Cognition – a Cybernetics approach", in: Garvin, P.L. (ed), *Cognition, a Multiple View*, Spartan, New York, pp. 331–384

Ashby, W.R. (1956) *An Introduction to Cybernetics*,. Wiley, New York

Ashby, W.R. (1960) *Design for a Brain*, 2nd edn. Chapman and Hall, London

Ashby, W.R. (1964) "The next ten years", in: Tou, J.T. amd Wilcox, R.H. (ed) *Computer and Information Sciences*, Spartan, Washington, pp. 2–11

Axelrod, R. (1984) *Evolution of Cooperation*, Basic Books, New York

Banerji, R.B. (1969) *Theory of Problem Solving*, Elsevier, New York

Barlow, H.B. (1959) "Sensory mechanisms, the reduction of redundancy, and intelligence", in: *Mechanisation of Thought Processes* (Proceedings of Conference in the National Physical Laboratory, Teddington), HMSO, London, pp. 537–574

Barto, A.G. (1990) "Connectionist learning for control: an overview", in: Miller, W.T., Sutton, R.S. and Werbos, P.J. (eds) *Neural Networks for Control*, MIT Press, Cambridge, Mass., pp. 5–58

Barto, A.G., Sutton, R.S. and Anderson, C.W. (1983) "Neuronlike adaptive elements that can solve difficult learning control problems", *IEEE Trans. on Systems, Man and Cybernetics*, vol. **SMC-13** , no. 5, pp. 834–846

Bateson, G. (1973) "Double Bind, 1969", in: Bateson, G. *Steps to an Ecology of Mind*, Paladin, London, pp. 242–249

Bayliss, L.E. (1966) *Living Control Systems*, EUP, London, pp. 68-73

Beer, S. (1959) *Cybernetics and Management*, English Universities Press, London

Beer, S. (1962) "Towards the cybernetic factory", in: von Foerster, H. and Zopf, G.W. (eds), *Principles of Self-Organization*, Pergamon, Oxford, pp. 25–89

Beer, S. (1975) *Platform for Change*, Wiley, London

Beer, S. (2004) "Ten pints of Beer", *Kybernetes*, vol. **33**, no. 3/4, pp. 828–942

Bellman, R. (1961) *Adaptive Control Processes: A Guided Tour*, Princeton Univ. Press, Princeton, New Jersey

Bellman, R. (1964) "Dynamic programming, learning, and adaptive processing", in: Tou, J.T. amd Wilcox, R.H. (eds) *Computer and Information Sciences,* Spartan, Washington, pp. 375–380

Benzinger, T.H. (1969) "Heat regulation homeostasis of central temperature in man", *Physiol. Rev.* vol. **49**, pp. 671–759

Berg, H.C. (1975) "How bacteria swim", *Scientific American*, vol. **233**, no. 2 (August), pp. 36–44

Bishop, M. (2002) "Dancing with pixies: strong artificial intelligence and panpsychism", in: Preston, J. and Bishop, M. (eds) *Views into the Chinese Room: New Essays on Searle and Artificial Intelligence*, Oxford University Press, pp. 360–378

Blakemore, C. (1977) *Mechanics of the Mind*, University Press, Cambridge

Blakemore, C. and Cooper, G.F. (1970) "Development of the brain depends on visual environment", *Nature*, vol. **228**, pp. 477–478

Boden, M. (1977) *Artificial Intelligence and Natural Man*, Basic Books, New York

Bowden, B.V. (1953) (ed) *Faster Than Thought*, Pitman, N.Y.

Brady, J.M.(1981) (ed) *Computer Vision*, North-Holland, Amsterdam

Brooks, R.A. (1989) "The whole iguana", in: Brady, M. (ed) *Robotics Science*, MIT Press, Cambridge, MA, pp. 432–456

Brooks, R. (1999) *Cambrian Intelligence: The Early History of the New AI*, MIT Press, Cambridge, Mass. (reviewed in *Kybernetes*, vol. **30**, 2001, no. 1, pp. 105–107)

Brown, G.S. (1969) *Laws of Form*, Allen and Unwin, London

Bryan-Jones, J. (1986) "Algorithms must be used with caution", *Computing Magazine*, Feb. 27, p. 26.

Byrne, W.L. (1970) (ed) *Molecular Approaches to Learning and Memory*, Academic Press, N.Y. (critically reviewed by S.P.R. Rose in: *Int. J. Man-Machine Studies*, vol. **3**, pp. 385–387, 1971)

Carbonell, J.G. (1984) "Learning by analogy: formulating and generalizing plans from past experience", in: Michalski, R.S., Carnonell, J.G. and Mitchell, T.M. (ed) *Machine Learning*, Springer, Berlin, pp. 137–161

Chapman, B.L.M. (1959) "A self-organizing classification system", *Cybernetica* (Namur), vol. **2**, no. 3, pp. 152–161

Chouraqui, E. (1985) "Construction of a model for reasoning by analogy", in: Steels, L. and Campbell, J.A. (ed) *Progress in Artificial Intelligence*, Ellis Horwood, Chichester, pp. 169–183

Chung, S-H. (1976) "Ineffective nerve terminals", *Nature*, vol. **261**, pp. 190–191

Churchland, P.S. (1986) *Neurophilosophy – Toward a Unified Science of the Mind/Brain*, MIT Press. Cambridge, MA

Conway, F. and Siegelman, J. (2005) *Dark Hero of the Information Age: In Search of Norbert Wiener, the Father of Cybernetics*, Basic Books, New York

Copeland, B.J. (2004) "Computable numbers: a guide", in: Copeland, B.J. (ed) *The Essential Turing: The ideas that gave birth to the computer age*, Oxford University Press, pp. 5–57

Cornew, R.W., Houk, J.C. and Stark, L. (1967) "Fine control in the human temperature regulation system", *J. theor. Biol.* vol. **16**, pp. 406–426

Corning, P.A. (2005) *Holistic Darwinism: Synergy, Cybernetics, and the Bioeconomics of Evolution*, Univ. of Chicago Press

Craik, K.J.W. (1942, reprinted 1952) *The Nature of Explanation*, University Press, Cambridge

Craik, K.J.W. (1947) "Theory of the human operator in control systems", *British Journal of Psychology*, vol.**38**, pp. 56–61 and 142–148

Darwin, C. (1897) *The Formation of Vegetable Mould through the Action of Worms*, John Murray, London

Dayhoff, J.E., Hameroff, S., Swenberg, C.E. and Lahoz-Beltra, R. (1993) "The Neuronal Cytoskeleton: A Complex System that Subserves Neural Learning", in: Pribram, K.H. (ed)

Rethinking Neural Networks: Quantum Fields and Biological Data, Lawrence Erlbaum, Hillsdale, New Jersey, Chapter 12, (pp. 389–441)

Dickinson, A. (1987) "Conditioning", in: Gregory, R.L. (ed) *The Oxford Companion to the Mind*, University Press, Oxford, pp. 159–160

Draper, N. and Smith, H. (1966) *Applied Regression Analysis*, Wiley, N.Y.

Dreyfus, S. (1961) "Dynamic programming", in: Ackoff, R.L. (ed) *Progress in Operations Research*, vol. 1, Wiley, Ch. 5, pp 211–242

Dubbey, J.M. (1978) *The Mathematical Work of Charles Babbage*, University Press, Cambridge

Fischler, M.A. and Firschein, O. (1987) *Intelligence: The Eye, The Brain and the Computer*, Addison-Wesley, Reading, Mass., p. 308

Foulkes, J.D. (1959) "A class of machines which determines statistical structure of a sequence of characters", in: *IRE Wescon Convention Record*, part 4, pp. 66–73

Fulton, J.F. (1949) *Physiology of the Nervous System*, 3rd edn., O.U.P., N.Y.

Donaldson, P.E.K. (1960) "Error decorrelation: a technique for matching a class of functions" *Proc. 3rd Conf. on Medical Electronics*, London, p. 173

Donaldson, P.E.K. (1964) "Error decorrelation studies on a human operator performing a balancing task" *Medical Electronics and Biological Engineering*, vol. 2, pp. 393–410

Draper, C.S. and Li, Y.T. (1951) *Principles of Optimalizing Control Systems and an Application to the Internal Combustion Engine*, ASME Publication

Fischler, M.A. and Firschein, O. (1987) *Intelligence: The Eye, the Brain and the Computer*, Addison-Wesley, Reading, MA

Friedberg, R.M. (1958) "A learning machine, part 1", *IBM Jnl. Res. and Dev.*, vol. 2, pp. 2–13

Gabor, D., Wilby, W.P.L. and Woodcock, R. (1961) "A universal non-linear filter, predictor and simulator which optimizes itself by a learning process" *Proc. I.E.E. (London)*, vol. B13, pp. 422–435

Gabor, D. (1968a) "Holographic model of temporal recall", *Nature*, vol. 217, p. 584

Gabor, D. (1968b) "Improved holographic model of temporal recall", *Nature*, vol. 217, pp. 1288–1289

Gaines, B.R. (1977) "System identification, approximation and complexity", *Int. J. General Systems*, vol. 3, pp. 145–174

Gardner, M. (1963) "A new paradox, and variations on it, about a man condemned to be hanged", *Scientific American*, vol. 208, no. 3, pp. 144–154

Garvey, J.E. (1963) (ed) *Self-Organizing Systems*, Office of Naval Research, Washington

George, F.H. (1962) "Minds, machines and Gödel: another reply to Mr Lucas", *Philosophy*, vol. 37, pp. 62–63

George, F.H. (1972) "Mechanism, interrogation and incompleteness", *Kybernetes*, vol. 1, no. 2, pp. 109–114

George, F.H. (1977) *The Foundations of Cybernetics*, Gordon and Breach, London, p. xiii

George, F.H. (1986) *Artificial Intelligence: Its Philosophy and Neural Context*, Gordon and Breach, N.Y.

Glushkov, V.M. (1966) *Introduction to Cybernetics*, Academic, New York, p. 140

Gross, C.G. (2002) "Genealogy of the 'Grandmother Cell'", *The Neuroscientist*, vol. 8, no. 5, pp. 512–518

Haldane, J.B.S. (1955) "Population genetics", *New Biology*, vol. 18, pp. 34–51

Hall, R.P. (1989) "Computational approaches to analogical reasoning: a comparative analysis", *Artificial Intelligence*, vol. 39, pp. 39–120

Hamilton, W.D. (1964) "The genetical evolution of social behaviour", *J. Theor. Biol.*, vol. 7, pp. 1–16 and 17–52

Hebb, D.O. (1949) *Organization of Behavior*, Wiley, New York, p. 62

Hebb, D.O. (1980) *Essay on Mind*, Lawrence Erlbaum, Hillsdale, N.J.

Hernandez-Peon, R., Scherrer, H. and Jouvet, M. (1956) "Modification of electrical activity in cochlear nucleus during 'attention' in unanaesthetized cat", *Science*, vol. 123, pp. 331–332

Hinton, G.E. (1989) "Connectionist learning procedures", *Artificial Intelligence*, vol. 40, pp. 185–234

References 131

Hinton, G.E. and McClelland, J.L. (1988) "Learning representations by recirculation", in: Anderson, D.Z. (ed) *Neural Information Processing Systems*, American Inst. of Physics, New York, pp. 358–366

Hinton, G.E. and Sejnowski, T.J. (1986) "Learning and relearning in Boltzmann machines", in *Parallel Distributed Processing: Explorations in the Microstructure of Cognition*, vol. **1**, edited by Rumelhart, D.E., McClelland, J.L. and the PDP Research Group, MIT Press, Cambridge, Mass., pp. 282–317

Hopfield, J.J. (1982) "Neural networks and physical systems with emergent collective computational properties" *Proc. Natl. Acad. Sci. USA*, vol. **79**, pp. 2554–2558

Horridge, G.A. (1968) *Interneurons – Their Origin, Growth and Plasticity*, Freeman, London, p. 357 and pp. 382–383

Hubel, D.H. and Wiesel, T.N. (1962) "Receptive fields, binocular interaction and functional architecture in the cat's visual cortex", *J. Physiol.*, vol. **160**, pp. 106–154

Hubel, D.H. and Wiesel, T.N. (1963) "Receptive fields in the striate cortex of very young, visually inexperienced, kittens", *J. Neurophysiol.*, vol. **26**, pp. 994–1002

Hubel, D.H. and Wiesel, T.N. (1968) "Receptive fields and functional architecture of monkey striate cortex", *J. Physiol.*, vol. **195**, pp. 215–243

Jantsch, E. (1980) *The Self-Organizing Universe*, Pergamon, Oxford

Jeffrey, A. (1969) *Mathematics for Engineers and Scientists*, Nelson, London, p. 41

Jibu, M. and Yasue, K. (1994) "Is brain a biological photonic computer with subneuronal quantum optical networks?", in: Trappl, R. (ed) *Cybernetics and Systems '94*, World Scientific, Singapore, pp. 763–770

Johnson, B.D. (2008) "The Cybernetics of Society: The Governance of Self and Civilization", Jurlandia Institute, Maine, downloaded from: www.jurlandia.org/cybsoc.htm

Kilmer, W.L., McCulloch, W.S. and Blum, J, (1968) "Some mechanisms for a theory of the reticular formation", in: Mesarovic, M.D. (ed) *Systems Theory and Biology*, Springer, N.Y., pp. 286–375

Kilmer, W.L., McCulloch, W.S. and Blum, J, (1969) "A model of the vertebrate central command system", *Int. J. Man-Machine Studies*, vol. **1**, pp. 279–309

Klir, G.J. and Elias, D. (2003) *Architecture of Systems Problem Solving*, 2nd ed., Kluwer/Plenum, New York, pp. 44–47

Klir, G.J. (2006) *Uncertainty and Information: Foundations of Generalized Information Theory*, Wiley, Hoboken

Kochen, M. (1981) "Appropriate approximation in concept genesis", in: Lasker, G.E. (ed) *Applied Systems and Cybernetics*, Pergamon, N.Y., pp. 613–618

Lawrence, S., 6Giles, C.L. and Fong, S. (2000) "Natural language grammatical inference with recurrent neural networks", *IEEE Trans. on Knowledge and Data Engineering*, vol. **12**, no. 1, pp. 126–140

le Cun, Y. (1985) "Une procédure d'apprentissage pour réseau a sequil asymétrique (A learning procedure for asymmetric threshold network)" *Proceedings of Cognitiva 85*, pp. 599–604 (quoted by Barto 1990)

le Cun, Y. (1988) "A theoretical framework for backpropagation", in: Touretzky, D., Hinton, G. and Sejnowski, T. (ed) Proceedings of the 1988 Connectionist Models Summer School, Morgan Kaufmann, San Mateo, Calif., pp. 21–28 (quoted by Barto 1990)

Lenat, D.B. (1984) "The role of heuristics in learning by discovery – three case studies", in: Michalski, R.S., Carbonell, J.G. and Mitchell, T.M. (ed) *Machine Learning*, Springer, Berlin, pp. 243–306

Lettvin, J.Y., Maturana, H.R., McCulloch, W.S. and Pitts, W.H. (1959) "What the frog's eye tells the frog's brain", Proc. I.R.E., vol. 47, pp. 1940–1951

Lettvin, J.Y., Maturana, H.R., Pitts, W.H. and McCulloch, W.S. (1961) "Two remarks on the visual system of the frog", in: Rosenblith, W.A. (ed) *Sensory Communication*, MIT Press, pp. 757–776

Lovelock, J.E. (1979) *Gaia: A New Look at Life on Earth*, Oxford University Press

Lovelock, J.E. (1983) "Gaia as seen through the atmosphere", in: Westbroek, P. and de Jong, E.W. (eds) *Biomineralization and Biological Metal Accumulation*, Reidel, Dordrecht, pp. 15–25

Lovelock, J.E. (1986) "Gaia: the world as a living organism", *New Scientist*, vol. **18**, December, pp. 25–28

Lovelock, James (2006) *The Revenge of Gaia: Why the Earth is Fighting Back – and How We Can Still Save Humanity*, with Foreword by Sir Crispin Tickell, Allen Lane, London

Lubbock, J.K. (1961) "A self-optimizing non-linear filter", *Proc. I.E.E. (London)*, vol. **B13**, pp. 439–440

Lucas, J.R. (1961) "Minds, machines and Gödel", *Philosophy*, vol. **36**, pp. 112–127

MacKay, D.M. (1959) "On the combination of digital and analogue techniques in the design of analytical engines", in: *Mechanisation of Thought Processes*, HMSO, London, pp. 55–65

Masani, P.R. (1990) *Norbert Wiener, 1894–1964*, Birkhäuser, Basel

Masani, P.R. (1992) "The illusion that man creates reality – a retrograde trend in the cybernetical movement", *Kybernetes*, vol. **21**, no. 4, pp. 11–24

Masani, P.R. (2001) "Three modern enemies of science: materialism, existentialism, constructivism", *Kybernetes*, vol. **30**, no. 3, pp. 278–294

McClelland, J.L. and Elman, J.L. (1986) "Interactive processes in speech perception: the TRACE model" in: *Parallel Distributed Processing: Explorations in the Microstructure of Cognition*, vol. **2**, edited by James L. McClelland, David E. Rumelhart and the PDP Research Group, MIT Press, Cambridge, Mass., pp. 58–121

McCorduck, P. (1979) *Machines Who Think*, Freeman, San Francisco

McCulloch, W.S. and Pitts, W. (1943) "A logical calculus of the ideas immanent in nervous activity", *Bull. Math. Biophyics*, vol. **5**, pp. 115–133

McCulloch, W.S. (1959) "Where is fancy bred?", private circulation from MIT; reprinted in: McCulloch, W.S., *Embodiments of Mind*, MIT Press, Cambridge, Mass., 1965, pp. 216–229

McCulloch, W.S. (1960) "What is a number, that a man may know it, and a man, that he may know a number?", *General Semantics Bulletin*, nos. 26 & 27, pp. 7–18; reprinted in: McCulloch, W.S., *Embodiments of Mind*, MIT Press, Cambridge, Mass., 1965, pp. 1–18

McCulloch, W.S. (1965) *Embodiments of Mind*, MIT Press, Cambridge, Mass.

Meditch, J.S. (1973) "A survey of data smoothing for linear and nonlinear dynamic systems", *Auutomatica*, vol. **9**, no. 5, pp. 151–162. Reprinted in Sorensen (1985)

Melzack, R. and Wall, P. (1988) *The Challenge of Pain*, 2nd edn., Penguin Books, London (1st edition 1982)

Michie, D. and Chambers, R.A. (1967) "Boxes: an experiment in adaptive control", in: *Machine Intelligence 2*, ed. E. Dale and D. Michie, Oliver and Boyd, Edinburgh, pp. 137–152

Mihram, D., Mihram, G.A. and Nowakowska, M. (1978) "The modern origins of the term 'cybernetics'", *Proc. 8th Int. Congress on Cybernetics*, Namur, Sept. 1976, pp. 411–423

Miller, K.D. (1998) "Ocular dominance and orientation columns", in: Arbib, M.A. (ed) *The Handbook of Brain Theory and Neural Networks*, MIT Press, pp. 660–665

Minsky, M.L. (1959a) "Some methods of artificial intelligence and heuristic programming", in: *Mechanisation of Thought Processes*, HMSO, London, pp. 3–36

Minsky, M.L. (1959b) Contribution to discussion, in: *Mechanisation of Thought Processes*, HMSO, London, p. 71

Minsky, M.L. (1963) "Steps toward artificial intelligence", in: Feigenbaum, E.A. and Feldman, J. (ed) *Computers and Thought*, McGraw-Hill, N.Y., pp. 406–450

Minsky, M.L. (1967) *Finite and Infinite Machines*, Prentice Hall, Englewood Cliffs, New Jersey

Minsky, M. (2003) "Why A.I. is brain-dead", *Wired Magazine*, issue 11.08, August (available from www.wired.com/wired/archive/11.08/view.html?pg=3)

Minsky, M.L. and Papert, S. (1969) *Perceptrons*, MIT Press, Cambridge, MA

Minsky, M. and Selfridge, O.G. (1961) "Learning in random nets", in: Cherry, E.C. (ed) *Information Theory*, Butterworth, London, pp. 335–347

Moore, W. (1989) *Schrödinger: Life and Thought*, Cambridge University Press, p. 308

Muir, A. (1958) "Computer sales forecasting", *Computer Jnl.* vol. **1**, p. 113

Nagel, E. and Newman, E.R. (1959) *Gödel's Proof*, Routledge and Kegan Paul, London

Nathan, P. (1969) *The Nervous System*, Penguin, Harmondsworth

Negoita, C.V. (1993) "On constructivism", *Kybernetes*, vol. **22**, no. 3, p. 63, and no. 7, p. 57

Newell, A., Shaw, J.C. and Simon, H.A. (1959) "Report on a general problem-solving program", in: *Proc. Int. Conf. on Information Processing*, UNESCO, Paris, pp. 256–264

Newell, A. and Simon, H. (1963) "GPS, a program that simulates human thought", in: *Computers and Thought*, ed. E.A. Feigenbaum and J. Feldman, McGraw-Hill, New York, pp. 279–293

Nguyen, D. and Widrow, B. (1990) "The truck backer-upper: An example of self-learning in neural networks", in: Miller, W.T., Sutton, R.S. and Werbos, P.J. (ed) *Neural Networks for Control*, MIT Press, Cambridge, Mass., pp. 287–299

Nilsson, N.J. (1965) *Learning Machines*, McGraw-Hill, N.Y.

Nilsson, N.J. (1971) *Problem-Solving Methods in Artificial Intelligence*, McGraw-Hill, N.Y.

North, J.D. (1952) "The human transfer function in servo systems", in Tustin, A. (main ed) *Automatic and Manual Control*, Butterworths, London, pp. 473–494

Orwell, G. (1989) *Nineteen Eighty-Four*, Penguin, Harmondsworth.

Ozinga, J.R. (1981) "Genetic altruism", *The Humanist*, vol. **41**, pp. 22–27

Papert, S. (1965) "Introduction", in: McCulloch, W.S., *Embodiments of Mind*, MIT Press, Cambridge, Mass., pp. xiii-xx

Parker, D.B. (1985) *Learning Logic*. Tech. Report TR-47, Massachusetts Institute of Technology, Cambridge, Mass. (quoted by Barto, 1990)

Partridge, D. (1986) *Artificial Intelligence: Applications in the Future of Software Engineering*, Ellis Horwood, Chichester, p. 173

Pask, G. (1962) "The logical type of illogical evolution", in: *Information Processing* (Proceedings of IFIP Congress), North-Holland, Amsterdam, p. 482

Pask, G. (1975) *Conversation, Cognition and Learning*, Elsevier, Amsterdam

Pedrycz, W. (1989) *Fuzzy Control and Fuzzy Systems*, Research Studies, Taunton, UK and Wiley, New York

Peitgen, H.-O., Jürgens, H. and Saupe, D. (1992) *Chaos and Fractals: New Frontiers of Science*, Springer, New York

Penrose, R., (1989) *Emperor's New Mind: Concerning Computers Minds and the Laws of Physics*, Oxford University Press, Oxford

Penrose, R. (2002) "Consciousness, Computation, and the Chinese Room", in: Preston, J. and Bishop, M. (ed) *Views into the Chinese Room: New Essays on Searle and Artificial Intelligence*, Oxford University Press, pp. 226–249

Poerksen, B. (2004) *The Certainty of Uncertainty: Dialogues Introducing Constructivism*, Imprint Academic, Exeter, UK

Popper, K.R. (1972) *The Logic of Scientific Discovery*, sixth impression, Hutchinson, London

Polya, G. (1954) *Induction and Analogy in Mathematics* (Vol. **1** of *Mathematics and Plausible Reasoning*) Princeton University Press, Princeton, New Jersey

Preston, J. and Bishop, M. (2002)(ed) *Views into the Chinese Room: New Essays on Searle and Artificial Intelligence*, Oxford University Press

Pribram, K.H. (1971) *Languages of the Brain: Experimental Paradoxes and Principles in Neurophysiology*, Prentice-Hall, Englewood Cliffs, N.J.

Pribram, Karl H. (1993) (ed) *Rethinking Neural Networks: Quantum Fields and Biological Data*, Lawrence Erlbaum, Hillsdale, New Jersey

Prigogine, I, and Stengers, I. (1984) *Order out of Chaos: Man's New Dialogue with Nature*, Fontana, London

Rapoport, A. (1982) "Cybernetics as a link between holistic and analytic theories of cognition" *Abstracts of Sixth European Meeting on Cybernetics and Systems Research (EMCSR)*, Vienna (appears by title only)

Roberts, L.G. (1960) "Pattern recognition with an adaptive network", in: *IRE Intl. Convention Record*, part 2, pp. 66–70

Rosen, R. (1985) *Anticipatory Systems*, Pergamon, Oxford

Rosenblatt, F. (1961) *Principles of Neurodynamics*, Spartan, N.Y.

Rosenbrock, H.H. (1960) "An automatic method for finding the greatest or least value of a function", *Computer Jnl.*, vol. 3, pp. 175–184

Rosenbrock, H.H. (1990) *Machines with a Purpose*, University Press, Oxford (reviewed in *Kybernetes*, vol. **21**, no. 2, pp. 69–71, 1992)

Rumelhart, D.E., Hinton, G.E. and Williams, R.J. (1986) "Learning internal representations by error propagation", in *Parallel Distributed Processing: Explorations in the Microstructure of Cognition*, vol. **1**, edited by Rumelhart, D.E., McClelland, J.L. and the PDP Research Group, MIT Press, Cambridge, Mass., pp. 318–362

Samuel, A.L. (1963) "Some studies in machine learning using the game of checkers", in: Feigenbaum, E.A. and Feldman, J. (ed) *Computers and Thought*, McGraw-Hill, New York, pp. 71–105

Scheibel, M. and Scheibel, A. (1968) "The brain stem – an integrative matrix", in: Mesarovic, M.D. (ed) *Systems Theory and Biology*, Springer, N.Y., pp. 261–285

Schwefel, H.-P. (1981) *Numerical Optimization of Computer Models*, Wiley, Chichester

Scriven, M. (1951) "Paradoxical announcements", *Mind*, vol. **60**, pp. 403–407

Searle, J. (1984) *Minds, Brains and Science* (The 1984 Reith Lectures), BBC, London

Sejnowski, T.J. (1986) "Open questions about computation in cerebral cortex" in: *Parallel Distributed Processing: Explorations in the Microstructure of Cognition*, vol. **2**, edited by James L. McClelland, David E. Rumelhart and the PDP Research Group, MIT Press, Cambridge, Mass., pp. 372–389

Selfridge, O.G. (1959) "Pandemonium: a paradigm for learning", in: *Mechanisation of Thought Processes*, HMSO, London, pp. 511–531

Selfridge, O.G. and Neisser, U. (1963) "Pattern recognition by machine", in: Feigenbaum, E.A. and Feldman, J. (ed) *Computers and Thought*, McGraw-Hill, N.Y., pp. 237–250

Shannon, C.E. and Weaver, W. (1949) *The Mathematical Theory of Communication*, Univ. of Illinois Press, Urbana

Sherrington, C. (1949) *Goethe on Nature and on Science*, Cambridge University Press

Singer, M. (1968) "Some quantitative aspects concerning the trophic role of the nerve cell" in: Mesarovic, M.D. (ed) *Systems Theory and Biology*, Springer, Berlin, pp. 233–245

Singer, W. (1990) "Self-organization of cognitive structures", in: Eccles, J. and Creutzfeld, O. (ed) *The Principles of Design and Operation of the Brain*, Springer, Berlin, pp. 119–135

Sommerhoff, G. (1950) *Analytical Biology*, Oxford University Press

Sommerhoff, G. (1974) *Logic of the Living Brain*, Wiley, London

Sorenson, H.W. (1985) (ed) *Kalman Filtering: Theory and Application*, IEEE Press, New York

Sorenson, H.W. and Sacks, J.E. (1971) "Recursive fading memory filtering", *Information Science*, vol. 3, pp. 101–119. Reprinted in Sorensen (1985)

Stafford, R.A. (1965) "A learning network model", in: Maxfield, M., Callahan, A. and Fogel, L.J. (ed) *Biophysics and Cybernetics Systems*, Spartan, Washington, pp. 81–87

Stark, L. (1959) "Stability, oscillations and noise in the pupil servomechanism", *Proc. I.R.E.*, vol. **47**, pp. 1925–1939

Sutton, R.S. (1990) "First results with Dyna, an integrated architecture for learning, planning and reacting", in: Miller, W.T., Sutton, R.S. and Werbos, P.J. (ed) *Neural Networks for Control*, MIT Press, Cambridge, Mass., pp. 179–189

Sutton, R.S. (1996) (ed) *Reinforcement Learning*, Kluwer, Boston (reprinted from special issue of Machine Learning, vol. **8**, nos. 3-4, 1992; reviewed in *Kybernetes* vol. **28**, 1999, no. 5, pp. 585–587)

Sutton, R.S. and Barto, A.G. (1998) *Reinforcement Learning: An Introduction*, MIT Press, Cambridge MA (reviewed in *Kybernetes* vol. **27**, 1998, no. 9, pp. 1093–1096)

Thom, R. (1983) *Mathematical Models of Morphogenesis*, Ellis Horwood, Chichester

Turing, A.M. (1950) "Computing machinery and intelligence", *Mind*, vol. **59**, pp. 433–460 (Reprinted in: Feigenbaum, E.A. and Feldman, J., *Computers and Thought*, McGraw-Hill 1963, pp. 11–35, and in: Copeland, B.J., *The Essential Turing: The ideas that gave birth to the computer age*, Oxford University Press 2002, pp. 441–464

Uhr, L. and Vossler, C. (1963) "A pattern-recognition program that generates evaluates, and adjusts its own parametwers", in: Feigenbaum, E.A. and Feldman, J. (ed) *Computers and Thought*, McGraw-Hill, New York, pp. 251–268

Uttley, A.M. (1956) "A theory of the mechanism of learning based on the computation of conditional probabilities", *Proc. First Int. Congress of Cybernetics, Namur*, Gauthier-Villars, Paris, pp. 830–856

Uttley, A.M. (1959) "Conditional probability computing in a nervous system", in: *Mechanisation of Thought Processes*, HMSO, London, pp. 119–147

Visser, H. and Molenaar, J. (1988) "Kalman filter analysis in dendroclimatology", *Biometrics,* vol. **44**, pp. 929–940

von Foerster, H. (1950) "Quantum mechanical theory of memory", in: von Foerster, H. (ed) *Cybernetics* (Transactions of the Sixth Conference, 1949), Josiah Macy Jr. Foundation, New York, pp. 112–145

von Foerster, H. and Poerksen, B. (2002) *Understanding Systems: Conversations on Epistemology and Ethics*, Kluwer/Plenum, New York (reference to memory mechanisms, pp. 133–135)

von Foerster, H. and Zopf, G.W. (1962) (ed) *Principles of Self-Organization*, Pergamon, Oxford

Wall, P.D. and Egger, M.D. (1971) "Formation of new connections in adult rat brains after partial deafferentation", *Nature*, vol. **232**, pp. 542–545

Wall, P.D., Fitzgerald, M. and Woolf, C.J. (1982) "Effects of capsaicin on receptive fields and on inhibitors in rat spinal cord", *Experimental Neurology*, vol. **78**, pp. 425–436

Walter, W.G. (1953) *The Living Brain*, Duckworth, London

Warwick, K. (1998) *In the Mind of the Machine*, Arrow Books, London

Weizenbaum, J. (1976) *Computer Power and Human Reason*, Freeman, San Francisco

Watson, A.J. and Lovelock, J.E. (1983) "Biological homeostasis of the global environment: the parable of Daisyworld", *Tellus* vol. **35B**, pp. 284–289

Werbos, P.J. (1974) Ph.D. results (below) attributed by Barto to this date.

Werbos, P.J. (1989) *Beyond Regression: New Tools for Prediction and Analysis in the Behavioral Sciences*, Ph.D. diss., Harvard Univ., Cambridge, Mass. (incorporated in Werbos, 1994)

Werbos, P.J. (1990a) "Overview of designs and capabilities", in: Miller, W.T., Sutton, R.S. and Werbos, P.J. (ed) *Neural Networks for Control*, MIT Press, Cambridge, Mass., pp. 59–66

Werbos, P.J. (1990b) "A menu of designs for reinforcement learning over time", in: Miller, W.T., Sutton, R.S. and Werbos, P.J. (ed) *Neural Networks for Control*, MIT Press, Cambridge, Mass., pp. 67–95

Werbos, P. (1992) "The cytoskeleton: why it may be crucial to human learning and neurocontrol". *Nanobiology*, vol. **1**, no. 1, pp. 75–95

Werbos, P.J. (1994) *The Roots of Backpropagation: From Ordered Derivatives to Neural Networks and Political Forecasting*, Wiley, New York

Whiteley, C.H. (1962) "Minds, machines and Gödel: a reply to Mr. Lucas", *Philosophy*, vol. **37**, pp. 61–62

Widrow, B. and Smith, F.W. (1964) "Pattern-recognizing control systems", in: Tou, J.T. and Wilcox, R.H. (ed) *Computer and Information Sciences*, Spartan, Washington, pp. 288–317

Wiener, N. (1948) *Cybernetics or Control and Communication in the Animal and the Machine*, Wiley, New York

Willshaw, D.J., Buneman, O.P and Longuett-Higgins, H.C. (1969) "Non-holographic associative memory", *Nature*, vol. **222**, pp. 960–962

Willshaw, D.J. (1981) "Holography, associative memory and inductive generalization", in: Hinton, G.E. and Anderson, J.A. (ed) *Parallel Models of Associative Memory*, Lawrence Erlbaum, Hillsdale, New Jersey, pp. 83–104

Winograd, T. (1973) *Understanding Natural Language*, Edinburgh University Press

Yovits, M.C. and Cameron, S. (1960) (ed) *Self-Organizing Systems: Proceedings of an Interdisciplinary Conference*, Pergamon, N.Y.

Yovits, M.C., Jacobi, J.T. and Goldstein, G.D. (1962) (ed) *Self-Organizing Systems 1962*, Spartan, Washington

Zadeh, L.A. (1965) "Fuzzy sets", *Information and Control*, vol. **8**, pp. 338–353

Zeeman, E.C. (1977) *Selected Papers 1972–77*, Addison-Wesley, Reading, Mass.

Index